basic
graphical
kinematics

SECOND EDITION

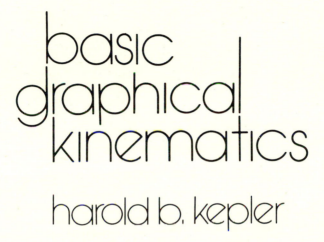

basic
graphical
kinematics

harold b. kepler

Associate Professor of Mechanical Engineering,
Air Force Institute of Technology

McGRAW-HILL BOOK COMPANY

New York	*Kuala Lumpur*	*Panama*
St. Louis	*London*	*Rio de Janeiro*
San Francisco	*Mexico*	*Singapore*
Düsseldorf	*Montreal*	*Sydney*
Johannesburg	*New Delhi*	*Toronto*

Library of Congress Cataloging in Publication Data
Kepler, Harold B.
  Basic graphical kinematics.

  Bibliography: p.
  1. Kinematics. 2. Graphical methods.
I. Title.
QA841.K4 1973 621.8′11 72–10890
ISBN 0–07–034171–0

KEPLER: BASIC GRAPHICAL KINEMATICS

8910 MUBP 8987654321

*The editors for this book were Robert Buchanan and
Cynthia Newby, the designer was Marsha Cohen,
and its production was supervised by Jim Lee.
It was printed by The Murray Printing Company and
bound by The Book Press, Inc.*

contents

preface

The emphasis of the second edition of this text remains on the graphical approach to kinematics. The graphical approach provides methods that are complementary with the methods usually presented in the physics and mechanics courses.

The level of this edition also remains essentially unchanged. This text is still considered to be a basic introduction to kinematics appropriate for the first or second year of an undergraduate engineering program or for the second year of a two-year technology program.

During the past decade, the first edition of this text was found to be quite effective and many favorable comments were received from both students and faculty members. Accordingly, there have been no wholesale changes incorporated in the second edition. Most of the changes are attempts to make the material more understandable or more teachable, and are the direct result of recurring student questions, student and faculty suggestions, and classroom experimentation.

The unusual features of this edition still include the following: (1) Vectors and vector equations are introduced early in a separate chapter so that their use may be studied without disrupting the continuity of later topics. (2) A chapter on kinematic drawing and displacements is included to familiarize students with graphical techniques and with the treatment of displacements in mechanisms. (3) Many illustrative examples are used, most of which are presented in a step-by-step fashion. (4) All four of the common graphical methods of obtaining velocities are presented, although the relative-velocity method is emphasized. (5) There are sufficient problems at the end of each chapter to permit much flexibility in problem assignments. Problems within each group are arranged, in general, in order of difficulty. (6) The velocity problems all have recommended velocity scales to ensure that the space requirements for the solutions are reasonable. (7) The acceleration problems have recommended locational data, space scales, velocity scales, and acceleration scales to make sure that the solutions will fit on $8\frac{1}{2}$- by 11-in. sheets. (8) The equivalent-linkage concept is presented, thereby simplifying the analysis of many mechanisms that would otherwise entail the troublesome Coriolis acceleration component. (9) An unusually simple, straightforward explanation of the Coriolis acceleration component makes it practical to include this topic in a one-quarter, first course in kinematics. (10) Methods are presented for developing complete motion curves (displacement, velocity, and acceleration) utilizing graphical calculus. (11) The chapter on cams not only provides the basic procedures for laying out cams to provide the conventional cam-follower motions but discusses the relative merits of the various types of motions. (12) The treatment of gears is unusually concise, yet complete enough to provide the student with techniques for analyzing or devising most gear-train arrangements. (13) Chapter 12 provides an extensive glimpse of some of the standard components or mechanisms available to machine designers and illustrates some of the classical methods of producing intermittent motion.

In addition to the large number of problems included in this text, there is a companion problem book (by the same author) available consisting of problems completely different from those in the text. A solutions manual is again available for the text problems and full-scale overlay solutions are available for the problem book.

The following is a summary of the most notable improvements in the second edition:

Chapter 2 has been expanded to give a more thorough treatment of the graphical solution of vector equations, so that the student will be better prepared to handle the velocity and acceleration problems in Chaps. 6 and 7. Also, many new problems have been added at the end of this chapter.

Chapter 3 has been expanded to include more proofs and developments to give the student a better feel for the application of these expressions in later chapters.

In Chap. 5, a derivation of the expression for the number of centros in a mechanism has been added, a new six-link centro example has been added including step-by-step centro diagrams, and most of the problems at the end of the chapter are entirely new.

In Chap. 6, the section on centro method for finding velocities has been expanded to clarify the two possible approaches: the general approach and the link-to-link approach. Also, an exhaustive six-link example has been added showing several alternative solutions. The four-link example illustrating the parallel-line method was expanded and a six-link example was added to more clearly indicate the versatility of this method. The explanation of the component method was expanded to include a more thorough explanation of the *line of proportion* as applied to floating links. The section on the relative-velocity method has been expanded by the addition of 2 six-link examples, one involving a trial-and-error technique not included in the first edition. There are 50 problems at the end of this chapter, most of which are entirely new.

In Chap.7, the order of the first three examples was changed to provide a smoother transition for the student. Both the explanation and the example problems pertaining to the Coriolis acceleration component have been simplified and a second rule has been added to assist the student in the formulation of problems involving the Coriolis component. Additionally, the Coriolis example (Example 7-6) and the equivalent-linkage example (Example 7-4) use the same problem to emphasize to the student that the results are the same. Nearly all of the problems at the end of this chapter are new.

In Chap. 8, a second method (chord method) for constructing motion curves is presented. Although the mirror method of constructing normals to a curve followed by constructing a perpendicular to each normal is probably the most accurate graphical method for constructing tangents to a curve, experience has shown that students obtain quicker, more consistent results using the less-accurate chord method.

Chapter 10 has been extensively rewritten to include the latest symbols, terms, and standard gear-tooth systems adopted by the American Gear Manufacturers' Association (AGMA) and the American Standards Association (ASA). Much more emphasis has been placed on such practical aspects of gearing as backlash, interference, undercutting, and contact ratio. New problems have been added embodying these practical considerations.

Chapter 11 has been updated to include the latest V-belt classifications. Examples involving pairs of gears and reverted gear trains include new, simplified alternative solutions which are more easily understood by many students.

Chapter 12 has been completely rewritten to incorporate many of the newer standard, stock mechanisms available to the machine designer.

As was pointed out in the Preface to the first edition, the author makes no particular claim for originality of material. The works of other authors have been heavily drawn upon. The books that were consulted at various times are listed in the bibliography. The sole justification for writing this book, then, lies in the particular topics covered, their sequence, and the particular manner in which they are presented.

The author wishes to express his appreciation to the many students who endured the aggravation associated with trying out the problems that are included in this text; to the reviewers of the manuscript who made many valuable suggestions; to the many manufacturers who provided photographs and problem material; and finally to my wife, Mary Ann, who spent so many hours in typing the manuscript and in proofreading.

Harold B. Kepler

basic
graphical
kinematics

intro-duction

1

1-1. Kinematics

Kinematics is a division of *mechanics*, which in turn is a division of *physics*. Mechanics deals with motion, masses, forces, and the effects of forces on bodies and is generally divided into two divisions: statics and dynamics. *Statics* deals with forces acting on rigid bodies at rest. *Dynamics* deals with motion and the effects of forces acting on rigid bodies in motion. Dynamics is divided into two parts: kinematics and kinetics. *Kinetics* is the study of forces acting on rigid bodies in motion and the effect of such forces in changing the motion. *Kinematics* is the study of motion without regard to the forces that cause it (see Fig. 1-1).

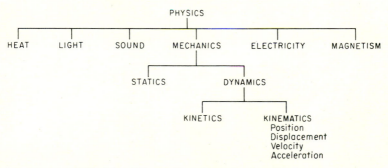

FIG. 1-1. Kinematics, a branch of mechanics.

When the principles of kinematics are applied to machines to determine the positions, displacements, velocities, and accelerations of their parts, the term *kinematics of machines* is used. In the kinematic analysis of machines, forces and forms of members may be disregarded. The various members of a machine are regarded as pure geometric forms.

1-2. Machine Design

The design of a machine is a complex procedure of which kinematics is just a phase. The general steps involved in machine design are as follows:

DETERMINING THE KINEMATIC SCHEME

This involves determining the motions necessary to accomplish the objective of the machine. At this stage only the general size relationships are established, and the various members are usually represented in skeleton form (kinematics).

DETERMINING THE FORCES INVOLVED

This involves the determination of the directions, magnitudes, and points of application of the external forces (statics and dynamics).

DETERMINING MEMBER PROPORTIONS AND MATERIALS

This involves deciding on the general shape, size, and material for each part so that it can successfully resist the forces involved (strength of materials).

DETAIL DESIGNING

This involves the determination of specific dimensions, tolerances, fits, manufacturing methods, lubrication methods, assembly methods, bearing types, fasteners, surface qualities, and protective coatings. Other considerations at this stage include safety features, cost of producing, customer appeal, ease of servicing, environmental conditions, vibration, and noise.

1-3. Terminology

LINK

In general, a link is any rigid body that is connected to other rigid bodies for the purpose of transmitting force or motion. Links include levers, rods, sliders, cranks, bellcranks, cams, gears, chains, belts, flexible couplings, universal joints, and the frames of machines. Belts and chains are included even though they are not rigid; their active portion (the portion in tension) is considered rigid for purposes of analysis. Actually, no link is completely rigid.

Links are so varied in form and function that it is difficult to classify them, but the following are some of the more common forms:

A *crank* is a link in the form of a rod or bar that executes complete rotations about a fixed center.

A *lever* (or rocker) is a link in the form of a rod or bar that oscillates through an angle, reversing its sense of rotation at certain intervals.

A *rocker arm* is a lever that is pivoted near the center, as shown in Fig. 1-2a.

A *bellcrank* is similar to a rocker arm except that it is bent at the pivot, as shown in Fig. 1-2b.

Gears and *cams* are also common links; they are discussed in detail in later chapters.

(a)   (b)

FIG. 1-2.   (a) Rocker arm. (b) Bellcrank.

Links are occasionally referred to as *simple links* if they have only two joints and as *complex links* if they have three or more joints or points to be analyzed.

KINEMATIC CHAIN

A kinematic chain is any group of links connected together for the purpose of transmitting forces or motions.

CONSTRAINED KINEMATIC CHAIN

This is an arrangement of two or more links such that a movement of one link causes a definite predictable movement of the other links (see Fig. 1-3a).

UNCONSTRAINED KINEMATIC CHAIN

This is an arrangement of links such that a movement of one link does *not* cause a definite predictable movement of the other links (see Fig. 1-3b).

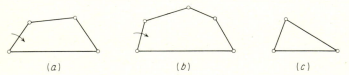

(a)   (b)   (c)

FIG. 1-3.   Kinematic chains. (a) Constrained. (b) Unconstrained. (c) Locked.

LOCKED KINEMATIC CHAIN

This type of chain is an arrangement of links such that no link can move relative to the other links in the chain, as shown in Fig. 1-3c. It is the basis of structural trusses.

MECHANISM

A mechanism is a kinematic chain with one link fixed. The fixed link is the *frame*, and the kinematic chain involved is almost always constrained.

MACHINE

A machine is a mechanism or group of mechanisms used to perform useful work. Its chief function is to adapt a source of power to some specific work requirement.

QUADRIC CHAIN

A quadric chain is any four-link kinematic chain.

1-4. Four-bar Mechanism

A four-bar mechanism is a mechanism having four rigid links with one link fixed. The fixed link is referred to as the *frame*; one of the rotating links is called the *driver* or *crank*; the other rotating link is called the *follower* or *rocker*; and the *floating link* (not rotating about a fixed center) is called the *connecting rod* or *coupler* (see Fig. 1-4*a*).

The fixed link may be *fixed absolutely* with respect to the earth or it may be *fixed* only *relatively* with respect to the other links, as in the case of a mechanism mounted in an airplane.

Depending upon the arrangement and proportions of the links, the two pivoted links may both rotate through 360°, one may rotate through 360° while the other oscillates, or both may oscillate. The mechanism may be *open* during part or all of its cycle, as shown in Fig. 1-4*a*, or it may be *crossed*, as shown in Fig. 1-4*b*.

If the four links form a parallelogram, as shown in Fig. 1-4*c*, the two pivoted links will both make complete revolutions, their

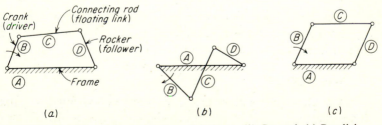

FIG. 1-4. Four-bar mechanisms. (*a*) Open. (*b*) Crossed. (*c*) Parallel.

angular velocities will be the same, and they will always be parallel to each other. There are two positions during the cycle, however, when the mechanism is not completely constrained. At these points, referred to as *dead points*, the follower could begin to rotate in a direction opposite to that of the driver.

Dead points occur in many mechanisms, but in most cases inertia, springs, or gravity prevents the undesired reversal or dead-center position. In Fig. 1-4a, for example, if the driver were rotated counterclockwise, a dead point would be reached when links C and D became collinear (lying in the same straight line).

1-5.   Drag-link Mechanism

The name *drag link* refers to a four-bar linkage in which both rotating links (cranks) make complete revolutions. The relative sizes of the links are such that the mechanism has no dead-point position. With reference to Fig. 1-5, the proportions of the links must be as follows: $C > A + D - B$ and $C < B + D - A$.

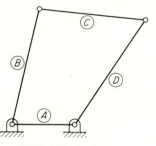

FIG. 1-5.   Drag link.

1-6.   Slider-crank Mechanism

The slider crank is a special case of the four-bar mechanism. As the follower in the regular four-bar mechanism in Fig. 1-4a gets longer, the path of the pin joint between the connecting rod and the follower approaches a straight line. In cases where straight-line motion of

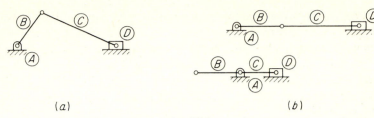

FIG. 1-6.   Slider crank.

this point is desirable, the follower link can be replaced with a slider, as shown in Fig. 1-6*a*.

Notice in Fig. 1-6*b* that the slider crank has two dead-center positions when links *B* and *C* are collinear. In applications where the slider is the driver, as in gasoline and steam engines, the machine is prevented from stopping on dead center either by a flywheel or by having two or more identical slider-crank arrangements out of phase with each other.

Figure 1-7 shows some variations of the slider crank. The scotch yoke is the equivalent of a slider crank with an infinitely long connecting rod; as a result, the slider reciprocates with simple harmonic motion and is useful as a sine-wave generator.

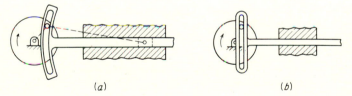

FIG. 1-7.   Variations of slider crank. (*a*) Regular. (*b*) Scotch yoke.

1-7. Quick-return Mechanisms

These mechanisms are used widely on machine tools to give a slow cutting stroke and a fast return stroke with a constant-angular-velocity drive crank. The ratio of the time required for the cutting stroke to the time required for the return stroke is called the *time ratio*. Figure 1-8 shows some of the common quick-return mechanisms.

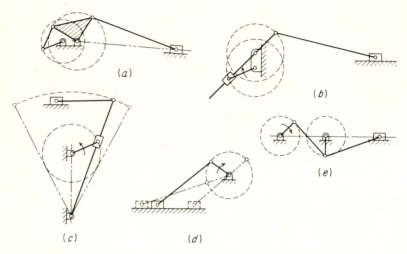

FIG. 1-8.   Quick-return mechanisms. (*a*) Drag link. (*b*) Whitworth. (*c*) Crank shaper. (*d*) Offset slider. (*e*) Crossed-link quick return.

1-8.   Toggle Mechanisms

The toggle mechanism is an adaptation of the slider crank. As the slider crank approaches one of its dead-center positions, the movement of the slider approaches zero and the ratio of the crank movement to the slider movement approaches infinity. This ratio of

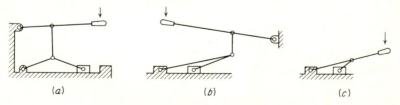

FIG. 1-9.   Toggle mechanisms.

movement is proportional to the mechanical advantage; thus, as the slider crank approaches dead center, the mechanical advantage, as between the crank and the slider, approaches infinity. The toggle mechanism consists of a basic slider-crank mechanism used in combination with another link or links or with one link extended, as shown in Fig. 1-9. Practical uses include punch presses, rock crushers, and toggle clamps. The clamp shown in Fig. 1-10 is a good example of a toggle mechanism. A force of 78 lb on the handle results in a 2,400-lb clamping force.

FIG. 1-10. Portable toggle clamp. (*Lapeer Manufacturing Company.*)

1-9. Straight-line Mechanisms

A straight-line mechanism is a mechanism that causes a point to travel in a straight line, or a nearly straight line, without being guided by a plane surface. These devices were very important in the early days of machinery before machine tools were available to generate smooth plane surfaces. When James Watt invented the steam engine, he had great need for a mechanism that would guide the joint between the piston rod and the beam that transmitted the piston motion to the crank.

Figure 1-11 shows some examples of straight-line mechanisms, where point *P* in each case describes approximately a straight line. The Scott-Russell mechanism in Fig. 1-11*a* is not actually a straight-line mechanism, for it utilizes a plane surface.

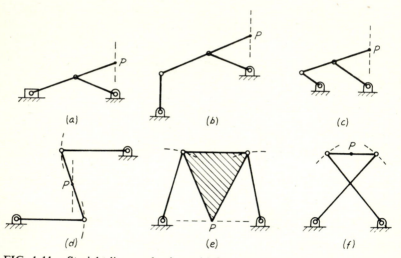

FIG. 1-11.   Straight-line mechanisms. (*a*) Scott-Russell. (*b*) and (*c*) Variations of Scott-Russell. (*d*) Watt's. (*e*) Robert's. (*f*) Tchebicheff's.

1-10.   Parallel Mechanisms

The drafting machine and the pantograph shown in Fig. 1-12 are familiar mechanisms that operate on the parallelogram principle. The drafting machine makes it possible to draw horizontal and vertical lines anywhere on the drawing board. The pantograph is used for duplicating, enlarging, or reducing figures as shown. A pencil or pen at *A* follows the motions of the stylus at *B* (the stylus and pen are interchangeable). This mechanism is used in many *copying* types of milling machines and engraving machines.

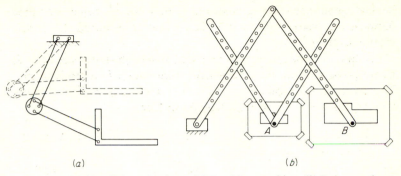

FIG. 1-12.   Parallel mechanisms. (*a*) Drafting machine. (*b*) Pantograph.

1-11.   Mechanism Inversions

An *inversion* of a mechanism may be defined as the choice of a different link in a mechanism to serve as the fixed link or frame. Thus in Fig. 1-13, any of the four links may be arbitrarily selected as the fixed link, and each arrangement is an inversion of the others.

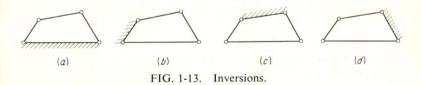

FIG. 1-13.   Inversions.

1-12.   Link Pairing

The connection of any two links is called a pair. The two links involved are called elements of the pair. Pairs are classified in two ways. (*a*) They may be classified as *turning pairs, sliding pairs, spherical pairs*, and *screw pairs*. The first two are the most frequently occurring types. (*b*) They may be classified as higher and lower pairs. *Higher pairing* consists of line or point contact, as in the case of a roller or ball rolling on a surface. *Lower pairing* consists of surface contact between two links, as in a pin joint or a slider.

1-13.   Phase, Cycle, and Period

The *phase* of a machine or mechanism comprises the relative positions occupied by the parts of a machine at any given instant.

The *cycle* of a machine or mechanism is its motion from the instant when its parts have certain relative positions to the instant when they assume again these same positions approaching them from the same directions. From an analysis standpoint, it matters little which phase is considered the beginning and end of the cycle.

The *period* of a machine or mechanism is the time required for one cycle.

vectors

2-1. Introduction

The basic quantities of mechanics are of two types: scalar and vector. *Scalar quantities* have magnitude only (like speed or angular velocity); *vector quantities* have both magnitude and direction. The elements of motion used in kinematic analysis (displacement, velocity, and acceleration) all have magnitude and direction and may, therefore, be expressed as vectors.

A vector is drawn as a straight line with an arrowhead at one end. Its length must be drawn to some particular scale if it is to have graphical meaning. The arrowhead shows the *sense* of the vector as differentiated from the *direction* or *line of action* of the vector. The plain end (tail) of a vector is its *origin*; the arrowhead end is its *terminus*.

Vectors that lie in one plane are called *coplanar*; vectors that do not lie in one plane (space vectors) are called *noncoplanar*. Since this text is concerned with plane motion, only coplanar vectors are discussed.[1]

In vector analysis it is shown that vectors can be manipulated in much the same way as algebraic quantities. That is, they can be added, subtracted, multiplied, divided, differentiated, etc. The only

[1] Graphical problems involving noncoplanar vectors require the use of descriptive geometry.

operations required in this text, however, are addition and subtraction. The symbols $\leftrightarrow$ and $\rightarrow$ are used to distinguish vector addition and subtraction from the algebraic operations.

2-2.   Vector Addition ($\leftrightarrow$)

Vectors are added by placing them head to tail while maintaining their correct directions and magnitudes. Figure 2-1*b* illustrates the addition of vectors *A* and *B*. Figure 2-1*c* shows how two vectors can also be added by placing them concurrent (tail to tail) and completing a parallelogram. This concept is useful in the resolution

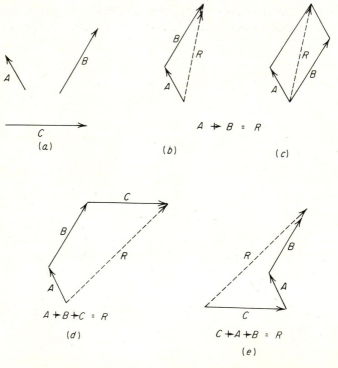

FIG. 2-1.   Vector addition.

of vectors discussed later. Figure 2-1*d* and *e* shows the addition of vectors *A*, *B*, and *C*. It should be observed that their sum is independent of their sequence.

2-3. Vector Subtraction (→)

The most straightforward method for subtracting one vector from another consists of adding the negative value (equal magnitude but opposite sense) of the vector to be subtracted. For example, in Fig. 2-2*b*, vector *B* is subtracted from vector *A* by adding *A* and (→*B*). A parallelogram is constructed in Fig. 2-2*c* to show that the same result can be achieved by placing vectors *A* and *B* concurrent and connecting their termini to obtain their difference *R*. Notice that the *difference vector points toward the vector being subtracted from.*

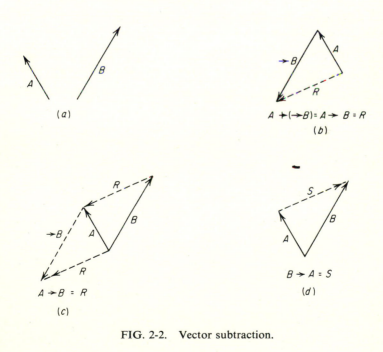

FIG. 2-2. Vector subtraction.

Figure 2-2*d* shows this *concurrent method of subtraction* used to obtain $B \rightarrow A = S$. Notice that $A \rightarrow B$ and $B \rightarrow A$ have the same magnitude but opposite sense (that is, $S = \rightarrow R$).

The concurrent method of subtracting vectors is the one that is most useful in Chaps. 6 and 7.

2-4. Vector Equations

As shown in Fig. 2-2, vector operations can be expressed in equation form. The terms (vectors) can be transposed by changing their signs, just as algebraic terms are transposed. The equation $A \rightarrow B = D$, for example, can be rearranged to $A = B \looparrowright D$. Furthermore, vector equations are not limited as to the number of vectors involved. The acceleration equations encountered in Chap. 7 typically involve 6 or 7 vectors.

Figure 2-3 illustrates the significance of a vector equation. If the two additions representing the two sides of the equation are started at a common origin *o* they must end at a common point such as *p*. The result is independent of the order of the vectors representing each side of the equation, as shown in Fig. 2-3*b*.

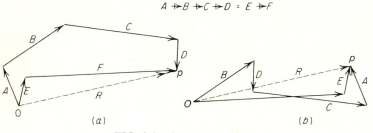

FIG. 2-3.   Vector equations.

Each vector in a vector equation represents two quantities, direction and magnitude. A vector equation can be solved if not more than two such quantities are unknown. For example, if one vector in an equation were completely unknown, the equation could be solved, or if one vector's direction were unknown while another vector's magnitude were unknown, the equation could be solved. Figure 2-4 shows the solution of a vector equation where two of the vectors have unknown magnitudes. Figure 2-4*c* and *d* shows how the known vectors for each side of the equation are placed head to

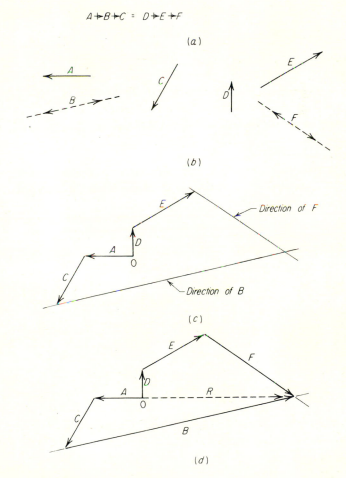

FIG. 2-4. Solution of vector equations—one unknown on each side.

tail starting from the common origin o. Since both sides of the equation must end up at the same point, the intersection of the two lines representing the directions of the two unknown vectors solves the equation and determines the terminus of R.

Figure 2-5 shows the solution of an equation where the two vectors whose magnitudes are unknown are on the same side of the equation. Notice in Fig. 2-5b that since the left side of the equation contains no unknowns, the sum R is known. It is obvious, then, that vectors D and E must close the gap between the terminus of vector C and the terminus of vector R. The equation is solved by drawing a line representing the direction of vector D through the terminus of vector C and a line representing the direction of vector E through the terminus of vector R, as shown in Fig. 2-5b. The intersection of these two lines produces the solution shown in Fig. 2-5c. An alternative placement of the two direction lines for vectors D and E is shown in Fig. 2-5d, producing the same results.

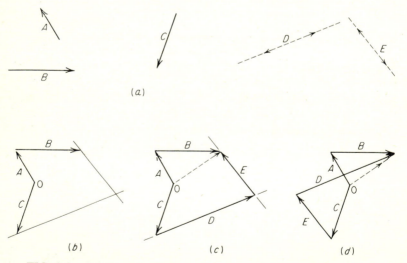

FIG. 2-5.   Solution of vector equations—two unknowns on one side.

In situations involving two unknown magnitudes there are usually two possible solutions and the context of the problem will dictate which of the two is of interest. Figure 2-6 illustrates this situation.

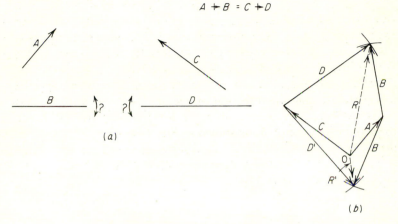

$$A + B = C + D$$

(a)

(b)

FIG. 2-6. Solution of vector equations—two possible solutions.

2-5. Composition of Coplanar Vectors

The composition of vectors is the same as the addition: it is the summation of two or more vectors. Any number of vectors can be added by simply placing them one after the other as was shown in Fig. 2-1*d* and *e*. The sum of the individual vectors (components) is called their *resultant*. The resultant is the equivalent of the individual vectors; it could replace them (see Fig. 2-7).

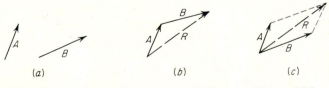

(a)   (b)   (c)

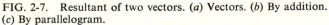

FIG. 2-7. Resultant of two vectors. (*a*) Vectors. (*b*) By addition. (*c*) By parallelogram.

A vector equal to the resultant in magnitude but opposite in sense is called the *equilibrant*. The equilibrant will exactly balance or offset the given system of vectors.

2-6.   Resolution of Coplanar Vectors

The resolution of a vector is the process of determining two or more (usually two) vectors that, when added together, will be the equivalent of the single vector. These individual vectors are called *components*. Obviously, there are an infinite number of pairs of components that could replace a vector, as shown in Fig. 2-8a. In order to resolve a vector into two particular components, it is necessary to know the directions of both components, the direction and magnitude of one, or the magnitudes of both. These conditions are illustrated in Fig. 2-8. If the components of a vector form a right angle, the components are called *normal* or *rectangular* components.

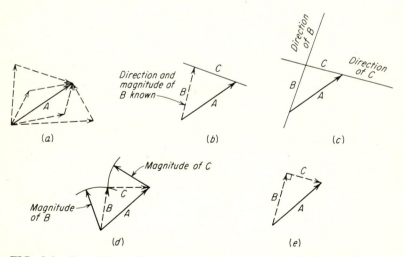

FIG. 2-8.   Resolution of vectors. (*a*) Infinite number of components. (*b*) Direction and magnitude of one component known. (*c*) Direction only of both components known. (*d*) Magnitudes only of both components known. (*e*) Rectangular components.

2-7. Orthogonal Components

An orthogonal component is obtained by projecting any vector upon a desired axis; it is the image or projection of the vector. In Fig. 2-9a, the orthogonal component of vector A on the x axis is shown. The result is a pair of rectangular components x and y. Figure 2-9b shows that there are many vectors like A_1, A_2, and A_3 that have the same orthogonal component x but that only one has a particular pair of rectangular components, such as x and y_2. Therefore, if the orthogonal component of a particular vector is given, the vector is not defined unless its direction is known or its other rectangular component is known. Orthogonal components are used in the component method of obtaining velocities, which is discussed in Art. 6-6.

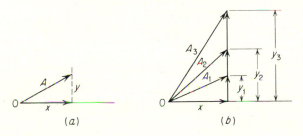

FIG. 2-9. Orthogonal components.

2-8. Vector Sense Notation

It is frequently necessary or desirable to describe a vector without drawing a picture of it. To do this it is necessary to be able to describe the vector's sense as well as its magnitude. There are several ways of doing this.

ANGLE

The sense of a vector may be described by giving the angle in degrees that it makes with the horizontal (x axis) measuring in the

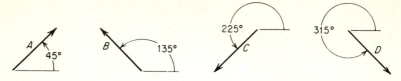

FIG. 2-10.   Vector sense—angle method.

conventional counterclockwise direction. Figure 2-10 shows several vectors and the angles that would be used to describe their senses.

BEARING

When vectors are involved in civil-engineering problems, their senses are usually described by their bearings. The bearing of a line is its east or west deviation in degrees from either the north or the south, whichever results in an angle less than 90°. Figure 2-11 shows four vectors and their bearings.

AZIMUTH BEARING

When vectors are involved in navigation problems, their senses are usually described by their azimuth bearings, or *headings*. The azimuth bearing of a line is its clockwise deviation in degrees from

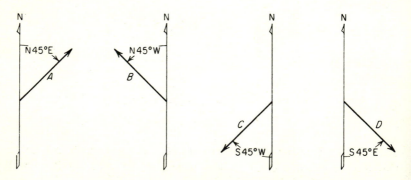

FIG. 2-11.   Vector sense—bearing method.

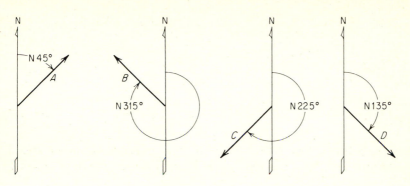

FIG. 2-12. Vector sense—azimuth bearing method.

the north. Figure 2-12 shows four vectors and their azimuth bearings. To avoid confusion, azimuth bearings must be preceded by N unless the words *heading* or *azimuth bearing* are used to distinguish them from angles measured from the *x* axis.

Problems

2-1. Determine graphically the sum of vectors A and B given in Fig. 2-13 and write the equation representing the operation.

FIG. 2-13. Problems 2-1 and 2-2.

2-2. Using the same vectors A and B given in Prob. 2-1, solve graphically the equation $R = A \rightarrow B$ using (*a*) the head-to-tail method shown in Fig. 2-2*b* and (*b*) the concurrent method shown in Fig. 2-2*d*.

2-3.   Solve graphically the vector equation shown in Fig. 2-14 by making two separate scale drawings using a different sequence of vectors each time.

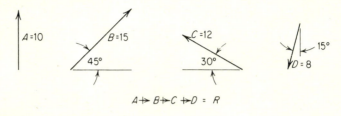

$$A \rightarrow B \rightarrow C \rightarrow D = R$$

FIG. 2-14.   Prob. 2-3.

2-4.   Solve graphically the vector equation $R = A \rightarrow B \rightarrow C \rightarrow D \rightarrow E$ using the vectors shown in Fig. 2-15.

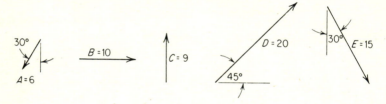

FIG. 2-15.   Prob. 2-4.

2-5.   Write the vector equation for the arrangement of vectors shown in Fig. 2-16.

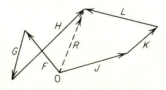

FIG. 2-16.   Prob. 2-5.

2-6. (*a*) Write the vector equation for the arrangement of vectors shown in Fig. 2-17. (*b*) Rewrite the equation to eliminate negative terms. (*c*) Draw the figure representing the new equation to verify that an equality still exists. Draw the vectors twice the size shown in Fig. 2-17.

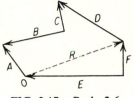

FIG. 2-17.   Prob. 2-6.

2-7.   Two vector equations can be written for vector *A*.

$$A = B \nrightarrow C \nrightarrow D \qquad \text{and} \qquad A = E \nrightarrow F$$

The directions for *B*, *C*, *D*, *E*, and *F* are known, and the magnitudes of *B*, *D*, and *F* are known, as shown in Fig. 2-18. Solve graphically the two vector equations simultaneously to determine *A*.

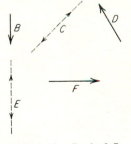

FIG. 2-18.   Prob. 2-7.

2-8.   (*a*) Resolve vector *A* into two rectangular components, one parallel to *mn* and one perpendicular to *mn*. (*b*) Resolve vector *A* into two rectangular components, one vertical and one horizontal. (*c*) Resolve vector *A* into two components, one along a 120° line and one along a 45° line. (*d*) Resolve vector *A* into two components, one along a 90° line and one 10 units long.

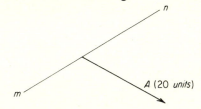

FIG. 2-19.   Prob. 2-8.

2-9.   Given two vectors *A* and *B* whose magnitudes are 6 and 10 units and whose senses are 60° and 15°, respectively, obtain graphically (*a*) their vector sum *C* and (*b*) their vector difference *D*.

2-10.   Resolve the vector *A* whose magnitude is 10 units and whose bearing is S30°W into two vectors, one along a N30°W–S30°E line and the other along a N45°E–S45°W line. Give the magnitudes of the two vectors. Show all three vectors with their origins at one point, i.e., concurrent.

2-11.   Resolve the vector *B* whose magnitude is 12 units and whose bearing is N15°E into two orthogonal components, one horizontal (*x* component) and one vertical (*y* component). Show the magnitudes of these components.

2-12.   Resolve the vector *C* whose magnitude is 8 units and whose azimuth bearing is N45° into two components, one of which has a magnitude of 4 units and an azimuth bearing of 0° (360°). Determine and label the magnitude and bearing of the second component.

2-13.   The velocities of two points *A* and *B* are 25 and 50 ft/sec, respectively. Their respective senses are 0° and 90°. It is shown in the the next chapter that the velocity of one moving point with respect to a second moving point is their vector difference. This is expressed as $V_{B/A} = V_B \rightarrow V_A$. Determine the velocity of *B* relative to *A* (both magnitude and sense).

2-14.   An airplane travels 12 mi due west, then 10 mi with a heading of 300°, and then makes a forced landing. In what direction and for what distance must a rescue helicopter fly to arrive at the scene of the landing?

2-15. An airplane travels 25 miles with a heading of 180°, then *s* miles with a heading of θ, arriving at a point P which is 30 miles from its starting point and which has a bearing of 240° with respect to the starting point. (*a*) Find *s* and θ. (*b*) Assign letters to the other vectors and write the vector equation for *s*.

2-16. Given the vector equation $J \rightarrow L = K$. If J is 20 in. long and has a sense of 75° and if L is 15 in. long and has a sense of 0°, what is the length and sense of K?

basic motion concepts

3

3-1. Motion

A body that is changing its position is said to be in motion. In the study of the kinematics of machines, the motion of a body must be described in relation to some other part of the machine under consideration, usually the frame. Motion described with respect to the frame is called *absolute motion* even though the frame itself is in motion. Motion described with respect to another moving link of the machine is called *relative motion*.

3-2. Plane Motion

A body is said to have *plane motion* when all points in the body move in parallel or coincident planes. Plane motion may consist of rotation, translation, or a combination of the two. This text treats the analysis of plane motion only; this is not a serious limitation, for most machine motions are either themselves plane motions or readily broken into two sets of plane motions.

ROTATION

A body is said to be rotating when all points in the body travel in circular paths, that is, remain at fixed distances from an *axis of rotation.*

TRANSLATION

When all points in a body have the same motion (that is, travel in parallel paths), the motion is said to be translation. If all of the points travel in *straight* parallel paths, the motion is *rectilinear translation*; if all of the points travel in *curved* parallel paths, the motion is *curvilinear translation.*

The locomotive-drive system shown in Fig. 3-1 illustrates the various plane motions. It is interesting to note that *any* plane motion of a body is completely described when the motions of any two points on the body are known. In the case of translation, the motion of only one point is required.

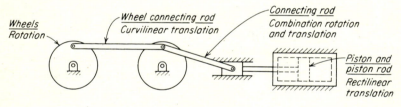

FIG. 3-1.   Examples of plane-motion types.

3-3.   Helical Motion

A body that is rotating about an axis as well as translating along the axis at a rate proportional to the rotation is said to have helical motion. Screw threads and worm gears are the most common examples of helical motion. Most spur gears are now made with helical teeth to provide smoother operation.

3-4.   Spherical Motion

A body that moves so that all its points remain at fixed distances from a common point is said to have spherical motion. This differs from rotation in that rotation is coplanar whereas spherical motion is noncoplanar, or three-dimensional. Rotation is a special case of spherical motion. A gear-shift lever on a pre-automatic-transmission automobile is a good example of spherical motion. A flyball governor on an engine is another example.

3-5. Motion of a Point

Since the motion of a rigid body is defined in terms of the motions of its points, it is necessary to consider first the motion of a single point.

PATH OF A POINT

A point moving from one position to another traces a line called a path. The *path* of a point is its *locus*.

DISPLACEMENT OF A POINT

A point moving from *A* to *B* along *any* path, no matter how devious, has a *change of position* of *AB* measured in a straight line, as shown in Fig. 3-2. This change of position is called the *displacement* of the point and is a *vector* quantity, since it has both direction and magnitude. The displacement of a point is discussed further in Art. 3-8.

It is important to notice that the total *distance* traveled by the point is the length of the curve *AB*. This distance is a *scalar* quantity and could be used to compute the *speed* of the point, which is also a scalar quantity.

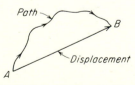

FIG. 3-2. Path and displacement.

3-6.   Combined Motions

If a point has made several *successive displacements*, its *total displacement* is the vector sum of its several displacements, as shown in Fig. 3-3. Furthermore, if a point has made several *simultaneous displacements*, its total displacement is exactly the same as if the several

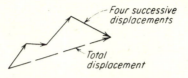

FIG. 3-3.   Successive displacements.

displacements had occurred separately. For example, if the airplane in Fig. 3-4 were to fly from *A* with a north heading and make no attempt to correct for the crosswind, its displacement would be *AC*, the sum of the two vectors involved (airplane velocity and wind velocity).

Therefore, a given displacement may, for purposes of analysis, be considered to be the combined effect of several displacements occurring simultaneously. This device of breaking displacements into components is extended to velocities and accelerations and becomes a valuable method of analysis.

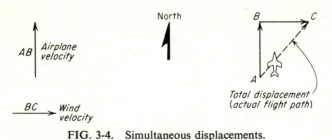

FIG. 3-4.   Simultaneous displacements.

3-7.   Position of a Point

The position of a point in a plane may be defined in two ways: (*a*) in terms of its distances from a pair of fixed rectangular axes, or (*b*) in terms of the length and direction of a line drawn to it from a single fixed point. The latter method is more convenient for discussions in this text.

The two positions B and B' of a moving point can be defined in terms of the lengths and directions of two lines drawn from some arbitrary reference point O. In Fig. 3-5, the position of B relative to O is represented by line OB, the position of B' relative to O is represented by line OB', and the position of B' relative to B is represented by line BB'. These position lines are vector quantities, having both direction and magnitude.

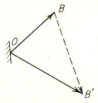

FIG. 3-5. Position and displacement of a moving point.

3-8. Displacement of a Point

As explained in Art. 3-5, the displacement of a point is its change in position and is independent of the actual path of the point. The displacement BB' in Fig. 3-5 can, therefore, be considered the vector change of the position of the moving point.

3-9. Velocity of a Point

The concept of velocity is introduced by considering the time required for a given displacement. The velocity of a moving point, then, is *the time rate of its change of position.*

3-10. Acceleration of a Point

The concept of acceleration is introduced by considering the time required for a given change in velocity. The acceleration of a moving point is *the time rate of its change of velocity.*

3-11. Velocity and Acceleration Relationships

It should be pointed out that the linear-motion expressions developed in this article involve vector quantities. The context of an actual problem would provide the directions for these vectors.

Instantaneous linear velocity is the time rate of change of linear displacement and is expressed as

$$V = \frac{ds}{dt} \approx \frac{\Delta s}{\Delta t} \qquad (3\text{-}1)$$

where V = linear velocity, ft/sec
s = linear displacement, ft
t = time, sec

Instantaneous linear acceleration is the time rate of change of linear velocity and is expressed as

$$a = \frac{dV}{dt} \approx \frac{\Delta V}{\Delta t}$$
$$= \frac{d^2 s}{dt^2} \qquad (3\text{-}2)$$

where a = linear acceleration, ft/sec².

Instantaneous angular velocity is the time rate of change of angular displacement and is expressed as

$$\omega = \frac{d\theta}{dt} \approx \frac{\Delta \theta}{\Delta t} \qquad (3\text{-}3)$$

where ω (omega) = angular velocity, rad/sec
θ (theta) = angular displacement, radians

Instantaneous angular acceleration is the time rate of change of angular velocity and is expressed as

$$\alpha = \frac{d\omega}{dt} \approx \frac{\Delta \omega}{\Delta t}$$
$$= \frac{d^2 \theta}{dt^2} \qquad (3\text{-}4)$$

where α (alpha) = angular acceleration, rad/sec².

RELATIONSHIP BETWEEN LINEAR AND ANGULAR VELOCITY

The instantaneous linear velocity of a point on a rotating body is proportional to its radius (see Fig. 3-6). This may be expressed as

$$V = r\omega \tag{3-5}$$

where V = linear velocity of point, ft/sec
$\quad r$ = radius of point, ft
$\quad \omega$ = angular velocity of body, rad/sec

PROOF

Starting with the definition of an angle as illustrated in Fig. 3-6*b*

$$d\theta = \frac{ds}{r}$$

or $\tag{a}$

$$ds = r\,d\theta$$

and the definitions of linear and angular velocities, Eqs. (3-1) and (3-3),

$$ds = V\,dt \tag{b}$$

and $\qquad d\theta = \omega\,dt \tag{c}$

then substituting Eqs. (*b*) and (*c*) into (*a*) gives the relationship

$$V\,dt = r\omega\,dt$$

or $\qquad V = r\omega$

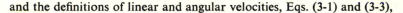

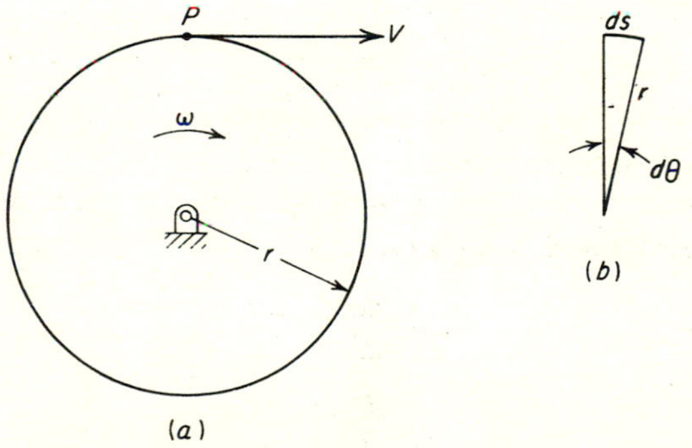

(*a*)

(*b*)

FIG. 3-6. Relationship between linear and angular velocity.

NORMAL AND TANGENTIAL ACCELERATIONS OF A ROTATING POINT

The instantaneous linear velocity of a point on a rotating body is easy to handle because its direction is known to be perpendicular to its radius. The instantaneous linear acceleration of a point on a rotating body is not so convenient because its direction is generally not obvious. For this reason, and for computational reasons, it is expedient to imagine that the linear acceleration of a point is made up of two components, one normal and one tangent to the path of the point. The acceleration of the point, then, is the vector sum of these two components as shown in Fig. 3-7a.

$$a = a^n + a^t \tag{3-6}$$

The *normal acceleration* is expressed as

$$a^n = V\omega = \frac{V^2}{r} = r\omega^2 \tag{3-7}$$

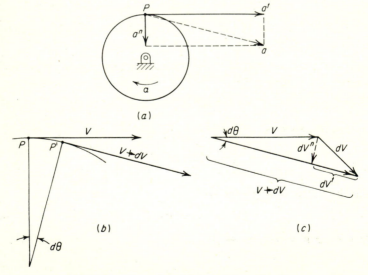

FIG. 3-7.   Normal and tangential accelerations.

where a^n = normal acceleration, ft/sec²
    V = linear velocity, ft/sec
    r = radius, ft
    ω = angular velocity, rad/sec

The *tangential acceleration*, which exists only if the body has an angular acceleration, is expressed as

$$a^t = r\alpha \qquad (3\text{-}8)$$

where a^t = tangential acceleration, ft/sec²
    r = radius, ft
    α = angular acceleration, rad/sec²

PROOF OF NORMAL ACCELERATION EXPRESSION

Starting with the definition of linear acceleration, $a = dV/dt$ [Eq. (3-2)], and extending its use to apply to the situation shown in Fig. 3-7c, the following two expressions may be written:

$$a^n = \frac{dV^n}{dt} \qquad (a)$$

$$a^t = \frac{dV^t}{dt} \qquad (b)$$

Also, it is evident from Fig. 3-7c that the definition of an angle would produce the expression

$$dV^n = V\,d\theta \qquad (c)$$

Then substituting Eq. (c) into Eq. (a), we obtain

$$a^n = \frac{dV^n}{dt} = \frac{V\,d\theta}{dt}$$

and since $d\theta/dt = \omega$ [Eq. (3-3)]

$$a^n = V\omega \qquad (d)$$

Also, since $V = r\omega$ [Eq. (3-5)]

$$a^n = V\omega = r\omega^2 = \frac{V^2}{r}$$

PROOF OF TANGENTIAL COMPONENT EXPRESSION

Starting with the expression $V = r\omega$ [Eq. (3-5)], adding the exponent t to be consistent with Fig. 3-7c,

$$V^t = r\omega \qquad (a)$$

and expressing this as a derivative with respect to time gives the expression

$$\frac{dV^t}{dt} = r\frac{d\omega}{dt} \qquad (b)$$

Then substituting $dV^t/dt = a^t$ [Eq. (3-2)] and $d\omega/dt = \alpha$ [Eq. (3-4)] into Eq. (b) results in the expression

$$a^t = r\alpha$$

3-12.   Relationship between Radians and Revolutions

In problems involving angular displacements, velocities, and accelerations, it is important that the angular data be in terms of radians rather than degrees or revolutions. A *radian* is an angle that subtends an arc equal to the radius of the arc. Therefore,

$$1 \text{ rev} = 2\pi \text{ radians} \qquad (3-9)$$

To convert from angular velocity expressed in rpm to angular velocity expressed in radians per second use the expression

$$\omega = \frac{2\pi n}{60} \qquad (3-10)$$

where ω = angular velocity, rad/sec
n = angular velocity, rpm

Equations (3-5) and (3-10) may be combined to produce the useful expression for the linear velocity of a point on a rotating body

$$V = \frac{2\pi r n}{60} \qquad (3-11)$$

where V = linear velocity, ft/sec
r = radius of point, ft
n = angular velocity of body, rpm

3-13. Summary of Practical Motion Relationships

Many practical problems involve initial and final velocities (V_0 and V) with constant acceleration. For these situations the linear acceleration can be expressed as[1]

$$a = \frac{V - V_0}{t} \tag{3-12}$$

This relationship may be rearranged as

$$V = V_0 + at \tag{3-13}$$

Displacement may be expressed as the product of the average velocity and the corresponding time interval or

$$s = \left(\frac{V_0 + V}{2}\right) t \tag{3-14}$$

Substituting Eq. (3-13) into Eq. (3-14) produces the expression

$$s = V_0 t + \tfrac{1}{2}at^2 \tag{3-15}$$

Equation (3-12) can be solved for t yielding the expression

$$t = \frac{V - V_0}{a} \tag{3-16}$$

Substituting Eq. (3-16) into Eq. (3-14) produces the expression

$$2as = V^2 - V_0^2 \tag{3-17}$$

A corresponding set of expressions can be developed for angular motion. Also, both the linear and the angular expressions simplify considerably when the initial velocity is zero. The following is a summary of both linear and angular relationships.

---

[1] When acceleration is constant, the *instantaneous* acceleration a is the same as the *average* acceleration $\bar{a}$.

When the initial velocity is not zero and the acceleration is constant:

Linear	*Angular*
$V = V_0 + at$	$\omega = \omega_0 + \alpha t$
$s = \frac{1}{2}(V_0 + V)t$	$\theta = \frac{1}{2}(\omega_0 + \omega)t$
$s = V_0 t + \frac{1}{2}at^2$	$\theta = \omega_0 t + \frac{1}{2}\alpha t^2$
$2as = V^2 - V_0^2$	$2\alpha\theta = \omega^2 - \omega_0^2$

When the initial velocity is zero and the acceleration is constant, the expressions simplify to the following:

Linear	*Angular*
$V = at$	$\omega = \alpha t$
$s = Vt$	$\theta = \omega t$
$s = \frac{1}{2}at^2$	$\theta = \frac{1}{2}\alpha t^2$
$2as = V^2$	$2\alpha\theta = \omega^2$

3-14.   Consistency of Units

The most common units for the above formulas have been indicated. If other units are used—inches instead of feet, minutes instead of seconds, or revolutions instead of radians—care must be exercised to ensure that the units within a particular expression are consistent. Since most dimensions on drawings are given in inches, a very common error is to use inches instead of feet in expressions involving radii. Occasionally, a student will even attempt to use degrees instead of radians.

3-15.   Relative Motion

Relative motion entails the concept of describing the motion of one point with respect to a second point that is itself moving.

RELATIVE DISPLACEMENT

In Art. 3-7 it was shown how the position of a point in a plane may be defined in terms of the length and direction of a line drawn to it from a single fixed point and how two positions B and B' of a moving point can be defined in terms of the lengths and directions of two lines drawn from some arbitrary fixed reference point. In Fig. 3-8 the position of B relative to O is represented by line OB and the position of B' is represented by OB'. Having both direction and magnitude, these two lines are vector quantities. The displacement s_B of the point B can be considered to be the vector change of its position with respect to a fixed point. Then $s_B = OB' \to OB$.

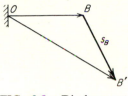

FIG. 3-8. Displacement of a point.

A moving point may also be defined in terms of another point that is moving instead of fixed. Consider the moving points A and B in Fig. 3-9. The original position of B relative to A is given by the line AB.

If the two points are given displacements s_A and s_B to positions A' and B', the position of B relative to A will be the line $A'B'$. It is easier to understand the relative change that takes place if the displacement of B to B' is thought of as two separate displacements—first to B'', then on to B'. Since the portion from B to B'' is parallel and equal to AA', the position of B relative to A remains unchanged. All the change in position of B relative to A, then, must take place during the second part of the displacement. Therefore, the displacement $s_{B/A}$ of B relative to A is represented by the line $B''B'$. It is evident from the figure that the absolute displacement s_B of B is equal

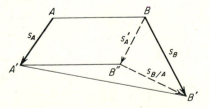

FIG. 3-9. Relative displacement.

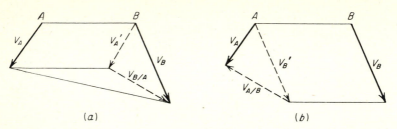

FIG. 3-10.   Relative velocity. (a) Velocity of B relative to A. (b) Velocity of A relative to B.

to the vector sum of the absolute displacement s_A of A and the displacement $s_{B/A}$ of B relative to A, which is expressed as

$$s_B = s_A \leftrightarrowtail s_{B/A}$$

or

$$s_{B/A} = s_B \rightarrow s_A$$

(3-18)

RELATIVE VELOCITY

If the displacements of points A and B in Fig. 3-9 are considered to take place during the same infinitesimal interval of time, then their velocity vectors will be proportional to their displacement vectors. Figure 3-10 shows the relative velocities of points A and B and is similar to Fig. 3-9. The relative velocity relationships may be expressed as

$$V_B = V_A \leftrightarrowtail V_{B/A} \qquad \text{or} \qquad V_{B/A} = V_B \rightarrow V_A$$

and

$$V_A = V_B \leftrightarrowtail V_{A/B} \qquad \text{or} \qquad V_{A/B} = V_A \rightarrow V_B$$

(3-19)

RELATIVE ACCELERATION

If the velocities of points A and B in Fig. 3-10 are regarded as changes in linear velocity (dV) that take place during the same infinitesimal interval of time (dt), then their acceleration vectors will be proportional to their velocity vectors $(a = dV/dt)$. Figure 3-11, then, showing the relative accelerations of points A and B, is similar to Figs. 3-9 and 3-10, and the relative acceleration relationships may be expressed as

$$a_B = a_A \leftrightarrowtail a_{B/A} \qquad \text{or} \qquad a_{B/A} = a_B \rightarrow a_A$$

and

$$a_A = a_B \leftrightarrowtail a_{A/B} \qquad \text{or} \qquad a_{A/B} = a_A \rightarrow a_B$$

(3-20)

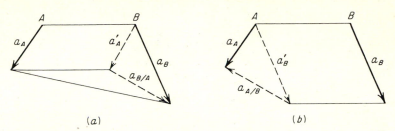

FIG. 3-11.   Relative acceleration. (*a*) Acceleration of *B* relative to *A*. (*b*) Acceleration of *A* relative to *B*.

SUMMARY OF RELATIVE MOTION

The relative motion of any two points is merely the vector difference of their absolute motions. Similarly, it can be shown that the relative angular motions of two rotating bodies may be defined as the difference of their absolute angular motions.[1]

Angular displacement: $\theta_{B/A} = \theta_B - \theta_A$ (3-21)

Angular velocity: $\omega_{B/A} = \omega_B - \omega_A$ (3-22)

Angular acceleration: $\alpha_{B/A} = \alpha_B - \alpha_A$ (3-23)

In the case of angular motions, however, the differences are algebraic (not vector) because angular motion is a scalar quantity.

If the linear velocity of one point on a link is known with respect to another point on the same link and if the distance between the two points is known, the angular velocity of the link can be determined by the expression $\omega = V/r$ [Eq. (3-5)].

In Fig. 3-12, the angular velocity of the rotating link can be found by using either of the absolute velocities V_A and V_B or by using their relative velocity $V_{A/B}$, which is their vector difference.

$$\omega = \frac{V_A}{r_A} = \frac{V_B}{r_B} = \frac{V_{A/B}}{r_{A/B}}$$

[1] Some convention must be adopted for positive and negative directions. It is suggested that clockwise (cw) be positive and counterclockwise (ccw) be negative.

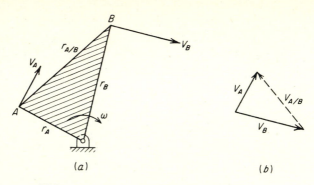

FIG. 3-12.   Angular velocity of a pivoted point.

EXAMPLE 3-1.   ANGULAR VELOCITY

In Fig. 3-13a, if the velocities of points A and B are known, the velocity of B relative to A can be obtained by the vector equation $V_{B/A} = V_B \rightarrow V_A$. The vector subtraction is shown in Fig. 3-13b. Then the angular velocity of the link is

$$\omega = \frac{V_{B/A}}{r_{B/A}} = \frac{5}{1.5} = 3.3 \text{ rad/sec (ccw)}$$

Similarly, if the tangential acceleration of one point on a link is known with respect to another point on the same link and if the distance between the two points is known, the angular acceleration of the link can be determined by the expression $\alpha = a^t/r$ [Eq. (3-8)].

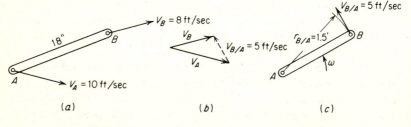

FIG. 3-13.   Angular velocity of a floating link.

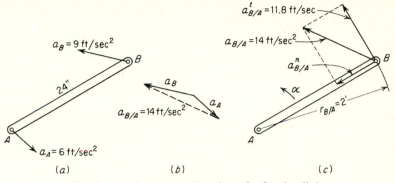

FIG. 3-14. Angular acceleration of a floating link.

EXAMPLE 3-2. ANGULAR ACCELERATION

In Fig. 3-14a, if the accelerations of points A and B are known, the acceleration of B relative to A can be found by the equation $a_{B/A} = a_B \rightarrow a_A$. This vector subtraction is shown in Fig. 3-14b. Figure 3-14c shows the relative acceleration resolved into the two components a^n and a^t. The angular acceleration of the link then is

$$\alpha = \frac{a^t_{B/A}}{r_{B/A}} = \frac{11.8}{2} = 5.9 \text{ rad/sec}^2 \text{ (ccw)}$$

Problems

3-1. Point A is on a slider that is accelerating uniformly along a straight path. The slider has a velocity of 10 ft/sec as it passes one point and a velocity of 30 ft/sec as it passes a second point 0.2 sec later. (a) What is the linear acceleration of A? (b) Through what distance has the slider traveled during this interval of time?

3-2. If a crank has an angular velocity of 30 rad/sec at one instant and has an angular velocity of 75 rad/sec 3 sec later, what is its acceleration? (Assume uniform acceleration.)

3-3. Lever A rotates about a shaft with an angular velocity of 50 rad/sec. Lever B rotates about the same shaft in the same direction with an angular velocity of 75 rad/sec. (a) What is the angular velocity of lever B with respect to lever A? (b) What would it be if lever B were rotating in the opposite direction?

3-4. A crank rotates at 100 rpm. (*a*) What is its angular velocity expressed in radians per second? (*b*) What is the linear velocity in feet per second of a point 6 in. from the center of rotation?

3-5. The linear velocities of two points on a crank are 80 and 50 ft/sec, respectively, while the crank rotates at 250 rpm. What is the distance between the two points?

3-6. A crank is 18 in. long and rotates uniformly at 150 rpm. (*a*) What is the linear velocity of point *A* at its extremity? (*b*) What is the acceleration of point *A*?

3-7. If the tip speed of a propeller 16 ft in diameter is not to exceed 1,100 ft/sec, what is the maximum engine rpm? (Assume direct drive.)

3-8. The recommended cutting speed for a particular steel is 600 surface feet per minute. At what rpm must the work rotate to give this speed if the diameter of the work piece is 6 in.?

3-9. Given two friction wheels in contact with each other. The driver is 6 in. in diameter and rotates at 955 rpm, and the follower (or driven wheel) is 16 in. in diameter. (*a*) Find the angular velocity of the driver in radians per second. (*b*) Find the tangential velocities of both rollers in feet per second. (*c*) Find the angular velocity of the follower in radians per second. (*d*) Find the velocity ratio ω_F/ω_D of the rollers.

The following problems require some graphical layout:

3-10. A 12-in.-diameter wheel rotating clockwise about a fixed center has an angular velocity of 5 rad/sec and is accelerating at 10 rad/sec². (*a*) Find the instantaneous linear velocity of a point *P* on the rim of the wheel. (*b*) Find the acceleration of point *P* (check graphically). *Scales:* space 1 in. = 4 in.; acceleration 1 in. = 4 ft/sec².

3-11. If a point *A* has a linear velocity of 10 ft/sec at 60° with respect to the *x* axis and if point *B* has a linear velocity of 12 ft/sec relative to *A* at 30° with respect to the *x* axis, what is the absolute linear velocity of *B*? *Scale:* 1 in. = 5 ft/sec.

3-12. Point *A* of the mechanism shown in the figure has an upward velocity of 12 ft/sec at the same time that point *B* has a velocity of 18 ft/sec. (*a*) Determine graphically the velocity of *B* relative to *A*. *Scale:* 1 in. = 10 ft/sec. (*b*) Calculate the angular velocity of the connecting link.

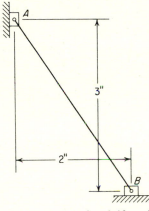

Fig. 3-15. Probs. 3-12 and 3-13.

3-13. If point *A* in the preceding problem has an upward acceleration of 100 ft/sec² at the same time that point *B* has an acceleration of 300 ft/sec² to the left, what is the angular acceleration of the connecting link? *Scale:* 1 in. = 100 ft/sec². *Hint:* Determine graphically the acceleration of *B* relative to *A*; then determine graphically the tangential component of this relative acceleration; then calculate the angular acceleration of the link.

3-14. An airplane is flying with a heading of N45° at an airspeed of 200 mph. There is a 50-mph wind from the west. What is the ground speed of the airplane, and what is its actual flight direction?

3-15. Airplane *A* is flying at a speed of 400 knots with a compass heading of N270°. Airplane *B* is flying at a speed of 500 knots with a heading of N45°. (*a*) What is the velocity of airplane *A* with respect to *B*? (*b*) What is the velocity of airplane *B* with respect to *A*?

3-16. Airplane *C* is flying at a speed of 230 mph with a compass heading of N60°. At a given instant, a second airplane *D* is located exactly 4 miles due east of airplane C, and its course and speed are N345° and 180 mph. How close will the airplanes be at their closest point?

3-17.   The center O of a 4-ft-diameter wheel rolling along the ground has a constant linear velocity of 10 ft/sec. At a given instant a point P on the rim of the wheel is located as shown in the figure. (*a*) What is the linear velocity of point P with respect to O? (*b*) What is the absolute velocity of P?

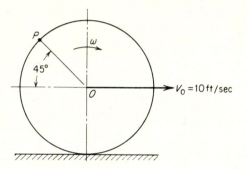

FIG. 3-16.   Probs. 3-17 and 3-18.

3-18.   If the center O of the rolling wheel in the preceding problem has an instantaneous linear acceleration of 15 ft/sec², what is the absolute acceleration of P? Solve graphically.

kinematic drawing and displacements

4-1. Kinematic Representation

In designing a mechanism, it is much easier to analyze or develop a kinematic scheme if the various parts of the mechanism are shown in their simplest form, using single lines where possible. In this way, drafting time is minimized, and the basic movements are not obscured by the complex configurations of the parts. Furthermore, in kinematic analysis, the mechanism must be studied in various phases, and this requires a new drawing for each phase. It would become quite time-consuming if the actual configurations were to be drawn for each phase.

Where an existing mechanism is to be analyzed, it is often difficult to look at an assembly drawing and recognize the actual geometry involved. This requires practice and is a skill that should be developed in kinematic analysis, whether analytical or graphical. Figure 4-1 shows examples of kinematic representation. Figure 4-1c shows a six-link mechanism[1] which produces large angular output oscillations.

[1] From Kurt Hain, *Applied Kinematics*, 2d ed., McGraw-Hill Book Company, New York, 1967.

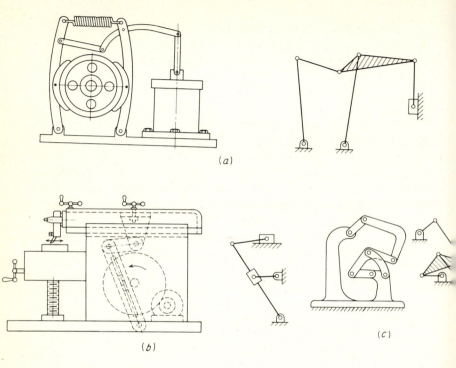

FIG. 4-1.   Examples of kinematic representation. (*a*) Magnetic safety brake. (*b*) Shaper. (*c*) Rocker linkage.

4-2.   Conventional Symbols

It is important to follow established conventions when depicting links, sliders, joints, ground marks, etc., so that the drawing will not be ambiguous.

A *pin joint* is usually drawn as a small circle approximately $\frac{3}{32}$ in. in diameter, as shown in Fig. 4-2*a*. Lines representing links usually do not extend inside the circle: a dot serves to locate the exact joint. For illustrative drawings or for drawings where extreme accuracy is not required, a simplified method for depicting a pin joint may be used consisting of a heavy dot, as shown in Fig. 4-2*b*.

A *simple link*, which is a link having only two joints, may usually be drawn as a single line, as shown in Fig. 4-2c.

A *complex link*, having three or more joints (or points to be analyzed), can be drawn as a solid figure bounded by straight lines and crosshatched as shown in Fig. 4-2d, or it can be made up of lines and arcs, as shown in Fig. 4-2e.

Through links are complex links that extend past a joint in such a way that they could be mistaken for two simple links. Figure 4-2f shows links A and B where link A extends past the joint. Notice that link A is drawn through the circle, whereas link B stops at the edge of the circle. Figure 4-2g shows an alternative way of handling this situation by making the through link a solid figure and crosshatching it. Figure 4-2h shows three separate links.

Pivots, or *fixed joints*, are represented as shown in Fig. 4-2i and j. It is important to realize that all fixed joints on a kinematic drawing

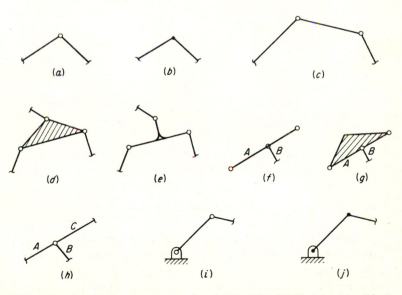

FIG. 4-2. Conventional representation. (*a*) Pin joint. (*b*) Pin joint (simplified). (*c*) Simple link. (*d*) and (*e*) Complex links. (*f*) and (*g*) Through links. (*h*) Three simple links. (*i*) Pivot (regular). (*j*) Pivot (simplified).

are part of the same fixed link or frame and are readily recognized by the *ground marks*, which are short lines resembling crosshatching.

A *slider* is a link that slides along another link and is usually drawn as shown in Fig. 4-3*a*. Its path may be curved, as shown in Fig. 4-3*b*; in this case the slider is the equivalent of the pivoted link shown in dashed lines. Alternative ways of drawing sliders are shown in Fig. 4-3*c*. Sliders may also move along other moving links as shown in Fig. 4-3*d*.

Pistons are usually shown as sliders except in the case where the cylinder pivots as shown in Fig. 4-3*e*.

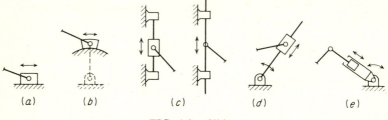

FIG. 4-3.   Sliders.

4-3.   Kinematic Design

As the new design of a mechanism evolves, it is the engineer's job to develop a kinematic scheme that will accomplish the desired mechanical movements. There are few concrete rules to guide him; he must rely largely on his experience with similar mechanisms. This phase of design requires much ingenuity, inventiveness, imagination, and patience.[1] A great amount of trial and error is necessary. This phase of kinematics cannot be treated in a basic kinematics course because of insufficient time. The scope of this text is restricted to the analysis of existing mechanisms, except for some gear and cam problems.

[1] Excellent sources of ideas for linkages are the following books: Franklin D. Jones (ed.), *Ingenious Mechanisms for Designers and Inventors*, The Industrial Press, New York, 1930; and Nicholas P. Chironis (ed.), *Machine Devices and Instrumentation*, McGraw-Hill Book Company, New York, 1966.

4-4. Kinematic Drawing

As stated earlier in this chapter, it is important to be able to recognize the basic geometry of a mechanism and to be able to reduce the mechanism to a simple skeleton drawing for purposes of kinematic analysis. Figure 4-4 illustrates the steps involved in converting an ordinary assembly drawing to a kinematic drawing.

Step 1. Draw *projectors* from the main points (usually joints) of the assembly drawing to an indefinite length.

Step 2. Establish a reference line in the assembly drawing perpendicular to the projectors. The location is arbitrary.

Step 3. Establish a second reference line at some convenient distance from the assembly drawing, again perpendicular to the projectors.

Step 4. Transfer distances, as shown in Fig. 4-4.

Step 5. Construct the skeleton drawing.

If the kinematic drawing is to be drawn to reduced scale, the direct projection of points and transferring of distances cannot be used.

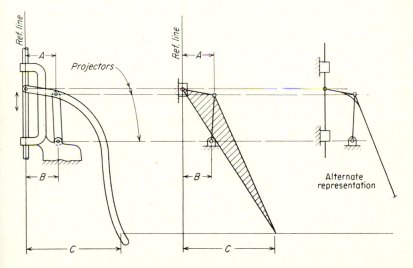

FIG. 4-4. Converting an assembly drawing to a kinematic drawing.

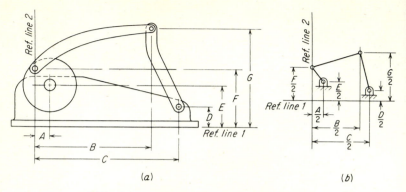

FIG. 4-5. Reduced-scale kinematic drawings.

The drawing must be constructed on two coordinate reference lines, as shown in Fig. 4-5.

4-5. Use of Drafting Equipment

Graphical kinematic analysis can be at least as accurate as analytical analysis in which a slide rule is used. Poor drafting techniques can cause inaccurate results just as poor slide-rule techniques can cause inaccurate analytical results.

PENCILS

In kinematic drawings, accuracy and neatness are more important than reproducibility. Therefore, the pencils used should be somewhat harder and sharper than those used in the preparation of working drawings. A 3H or 4H pencil is usually satisfactory. A softer pencil such as an H or 2H may be used for arrowheads and lettering.

TRIANGLES

The two most frequently occurring constructions in graphical kinematics are (*a*) to draw a line parallel to a given line and (*b*) to draw a line perpendicular to a given line. These constructions are easily accomplished with two triangles, as shown in Fig. 4-6.

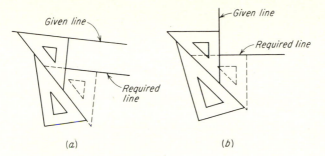

FIG. 4-6. Drawing parallel and perpendicular lines.

SCALES

The most convenient scale to use for kinematic drawing is the *engineer's scale*, which contains six scales: 10, 20, 30, 40, 50, and 60. The 10 scale is full size and has 10 divisions per inch. It is used for drawing to such scales as 1 in. = 10 in., 1 in. = 100 in., 1 in. = 10 ft/sec, or for full scale. The 20 scale has two main divisons per inch and 20 subdivisions per inch. It is used for drawing to such scales as 1 in. = 2 in., 1 in. = 20 in., 1 in. = 200 ft, or 1 in. = 2,000 ft/sec. In using an engineer's scale, one shows all data in decimals.

If the scale of a drawing is given in fractional form, such as $\frac{3}{8}$ in. = 1 ft or $\frac{3}{4}$ in. = 1 in., an architect's scale must be used. An *architect's scale* has one full-size scale graduated in sixteenths of an inch and several other scales for enlarged or reduced drawings. This scale can be used for dimensions given in feet and inches, such as 3 ft 6 in., or for dimensions given in inches and fractions of inches, such as $3\frac{3}{4}$ in. All scales, except the full-size scale, are *open-divided*, which means that each scale division represents a full unit (either a foot or an inch) with no intermediate scale divisions. Each scale has only one subdivided unit, which is located adjacent to the scale index (zero). It is important to realize that these subdivided portions are divided into 12 parts, corresponding to 12 in. to a foot. This is convenient for "feet-and-inches" dimensions, but is most inconvenient for "inches-and-fractions-of-inches" dimensions, since the fractions commonly used are fourths, eighths, sixteenths, thirty-seconds, and sixty-fourths. Therefore, any fraction with a denominator larger than 4 must be estimated.

TEMPLATES

Unimportant features of kinematic drawings such as small circles, sliders, etc., may be drawn quickly with plastic templates. If much of this type of drawing is anticipated, it is well worth the time to make a special template similar to the one shown in Fig. 4-7. The edge of the template can be used to draw links. There are many standard templates available for drawing small circles.

FIG. 4-7.   Template for making kinematic drawings.

4-6.   Useful Constructions

ANGLES

Where extreme accuracy is not required, angles may be drawn with a simple protractor. For more precise work, a special adjustable protractor with a vernier or a drafting machine can be used. Accurate angles can also be drawn by the *tangent method*, which consists in laying out a small right triangle with the *adjacent* side having 10 units and the *opposite* side having a number of units corresponding to the tangent of the angle, as shown in Fig. 4-8. Angles can also be

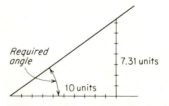

FIG. 4-8.   Tangent method of drawing and measuring angles. Required angle = 36°10′; tan 36°10′ = 0.731.

measured by this method. The units used are arbitrary, but the triangle should be as large as possible.

COPYING GEOMETRIC FIGURES

In drawing mechanisms in various positions, it is frequently necessary to redraw links. Links that rotate about fixed centers can be relocated by drawing arcs through the main points and rotating each point through the same angle, as shown in Fig. 4-9a. In most cases, however, it is easier to relocate one line of the link and then reconstruct the geometric form, using this one line as a base line. In the case of links not rotating about fixed centers, this is the only method possible. In Fig. 4-9b, the line AB of the link is relocated in

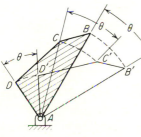

(a)

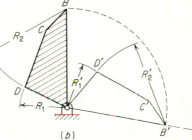

(b)

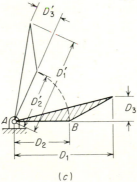

(c)

FIG. 4-9. Copying geometric figures.

the desired position. Points *C* and *D* are located by striking arcs from points *A* and *B*. In Fig. 4-9*c*, line *AB* again serves as a base line, but in this case offset measurements are used.

DRAWING NORMALS AND TANGENTS TO CURVES

The simplest and most accurate method of drawing a normal to a curve at a particular point consists in holding a mirror or polished surface (such as an injector-type razor blade) perpendicular to the paper as shown in Fig. 4-10. When the mirror is adjusted so that the *reflection* of the curve is a smooth continuation of the *visible* portion of the curve, the edge of the mirror is normal to the curve. The tangent is perpendicular to the normal and can be quickly drawn by the method shown in Fig. 4-6*b*.

Where a series of tangents is required, the *chord method* of drawing tangents is somewhat easier. This method is illustrated in Art. 8-3.

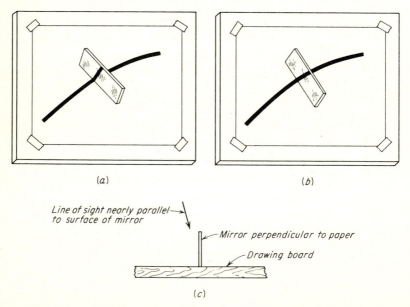

FIG. 4-10.   Drawing a normal to a curve. (*a*) Not normal. (*b*) Normal.

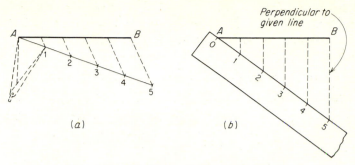

FIG. 4-11.   Dividing a line into equal parts.

DIVIDING A LINE

A line can be divided into any number of equal parts by either of the methods shown in Fig. 4-11. In Fig. 4-11a, a construction line is drawn from one end of the given line at some convenient angle. Along this line, the required number of spaces (five in this case) is stepped off with dividers, and the end point 5 connected with point B. The intermediate points on the given line are obtained by drawing lines parallel to line B5 through points 4, 3, 2, and 1.

In Fig. 4-11b, a construction line is first drawn through one end of the given line and perpendicular to the line. A scale is then placed diagonally as shown so that the number of scale divisions equals the number of segments desired. Marks are made at each scale division and these marks are extended straight up to the line.

The methods shown in Fig. 4-11 may be extended to divide a line into proportional parts. Figure 4-12 shows a line being divided

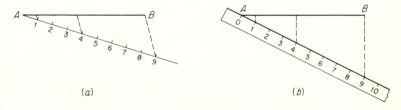

FIG. 4-12.   Dividing a line into proportional parts.

into parts proportional to 1, 3, and 5. This construction is useful in cam design.

4-7.   Displacement Drawing

In complete analysis work, it is necessary to draw a given mechanism in various phases of its cycle to determine the displacements, velocities, and accelerations of the various links in these various phases. It is necessary, therefore, to develop some skill at drawing mechanisms in various phases and determining limiting positions in the case of reciprocating mechanisms.

The slider-crank mechanism in Fig. 4-13*a* is arbitrarily shown in a given position. Figure 4-13*b* shows the mechanism for every 30°, which is sufficient in most cases. If the 30° positions do not coincide

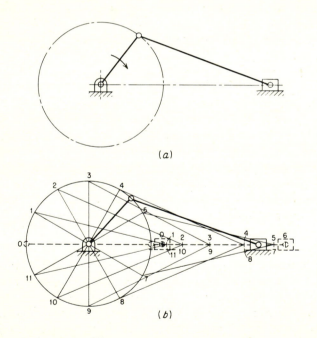

(*a*)

(*b*)

FIG. 4-13.   Displacement.

with dead-center or limit positions, these should be drawn in addition. In the figure, the increments were oriented so that they coincided with the limits of the slider stroke.

To avoid a confusing mass of lines, all positions are shown with light lines except extreme positions or *home* positions (at-rest positions for mechanisms that cycle once and stop), which are shown very heavy or in color. The various points are identified with small letters or numbers.

4-8. Displacement Diagrams

A displacement diagram is a *time-displacement plot* of some point or link in a mechanism with respect to the driver (or crank). The *time* usually consists of one revolution of the crank laid out horizontally along the x axis. The *displacement* of the follower is laid out vertically along the y axis. The units for the displacement may be linear or angular, depending upon the motion of the follower being analyzed.

A displacement curve provides a good graphic picture of the motion of a particular point or link during a complete cycle of the mechanism. It is possible to look at the curve and tell where peak velocities and accelerations occur (the velocity is proportional to the slope, and the acceleration is inversely proportional to the radius of curvature).

Figure 4-14 shows a displacement diagram in which the linear displacement of the slider has been plotted for one revolution of the crank. The peak velocities occur between points 3 and 4 and between points 9 and 10 (maximum slope), and the peak accelerations occur between points 6 and 7 and between points 11 and 1 (smallest radius of curvature). If the angular velocity of the crank were given, the horizontal scale could be converted to units of time, and the approximate linear velocity of the slider could be obtained for any position by determining the slope of the curve at that point.

Figure 4-15 shows a displacement diagram in which the angular displacement of the follower (rocker) has been plotted for one revolution of the crank. Again it is possible to spot the peak velocity and

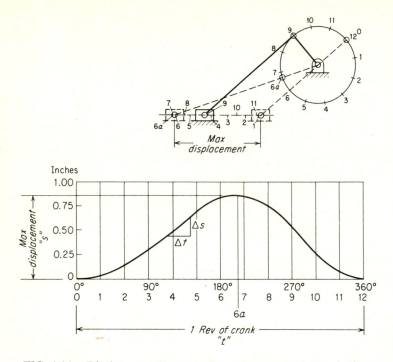

FIG. 4-14.   Displacement diagram—linear displacement. $V \approx \Delta s/\Delta t$.

acceleration points, and, if the angular velocity of the crank were known, the approximate angular velocity of the follower could be obtained for any position by determining the slope of the curve at that point. The vertical scale in this case would have to be converted to radians.

The determination of velocities and accelerations in mechanisms is treated in much detail in later chapters. Chapters 6 and 7 deal with methods for finding the exact velocity and acceleration of any point in a mechanism at a particular phase of the mechanism. Chapter 8 deals with the construction of all three of the motion curves—displacement, velocity, and acceleration.

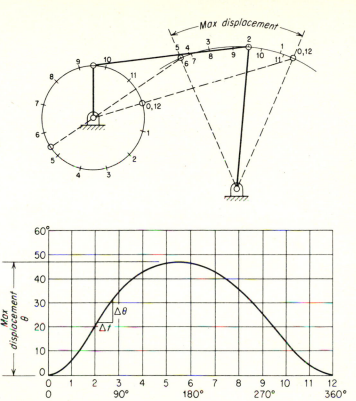

FIG. 4-15.   Displacement diagram—angular displacement. $\omega \approx \Delta\theta/\Delta t$.

Problems

4-1 to 4-6.   First make a kinematic drawing of the mechanism in the position shown, transferring distances directly from the book. Next, make a superimposed drawing of the mechanism in the open position. Clarity can be improved by using a colored pencil for the open position.

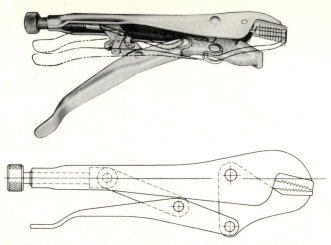

FIG. 4-16. Prob. 4-1. (*Lapeer Manufacturing Company.*)

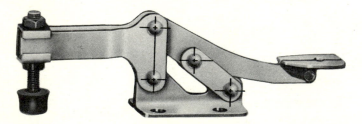

FIG. 4-17. Prob. 4-2. (*Lapeer Manufacturing Company.*)

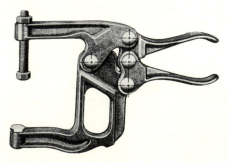

FIG. 4-18. Prob. 4-3. (*Lapeer Manufacturing Company.*)

FIG. 4-19. Prob. 4-4. (*Lapeer Manufacturing Company.*)

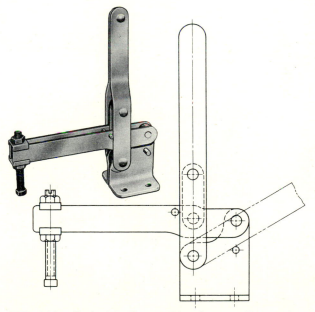

FIG. 4-20. Prob. 4-5. (*Lapeer Manufacturing Company.*)

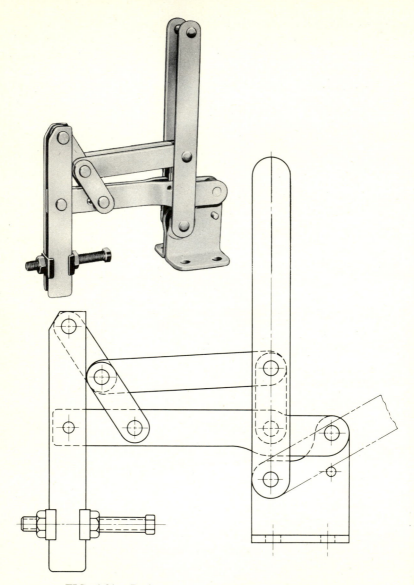

FIG. 4-21.   Prob. 4-6. (*Lapeer Manufacturing Company.*)

4-7. (*a*) Draw the mechanism in the position shown in Fig. 4-22. (*b*) Draw the two positions of the mechanism where link 4 is at its extreme left- and right-hand positions. (*c*) Plot the angular displacement (in degrees) of link 4 for the various positions of crank 2. Plot a point for every 30° of rotation of crank 2 with the 0° position corresponding to the extreme left-hand position of link 4.

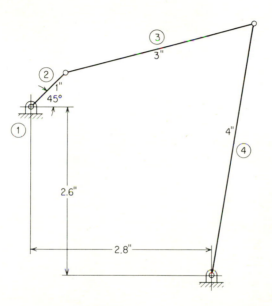

FIG. 4-22.   Prob. 4-7.

4-8. (*a*) Draw the mechanism (Fig. 4-23) in the position shown. (*b*) Draw the mechanism in the two extreme positions corresponding to the extreme left- and right-hand positions of the slider. (*c*) Plot the linear displacement of the slider for the various positions of crank 2 with the 0° position corresponding to the extreme left-hand slider position.

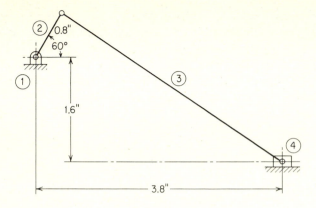

FIG. 4-23.   Prob. 4-8.

4-9. (*a*) Draw the mechanism (Fig. 4-24, Watt's) in the position shown. (*b*) Draw the phase of the mechanism where link 4 is rotated to its extreme clockwise position. (*c*) Draw the phase of the mechanism where link 2 is rotated to its extreme clockwise position. (*d*) Draw the path of the midpoint of link 3 for the positions between these extremes, obtaining at least eight points.

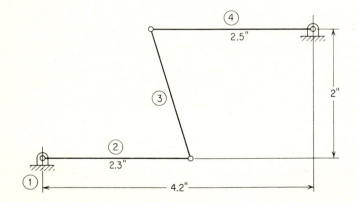

FIG. 4-24.   Prob. 4-9.

4-10. (*a*) Draw the mechanism (Fig. 4-25, Robert's) in the position shown. (*b*) Draw the phase of the mechanism where link 2 is rotated to its extreme counterclockwise position. (*c*) Draw the phase of the mechanism where link 4 is rotated to its extreme clockwise position. (*d*) Draw the path of point *A* for the positions between these extremes, obtaining at least eight points.

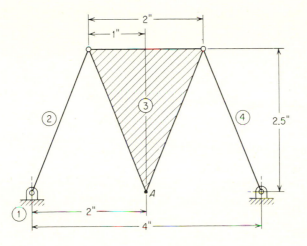

FIG. 4-25.   Prob. 4-10.

4-11. (*a*) Draw the mechanism (Fig. 4-26, Scott-Russell) in the position shown. (*b*) Draw the phase of the mechanism where link 3 is 60° above horizontal. (*c*) Draw the phase of the mechanism where link 3 is 60° below horizontal. (*d*) Draw the path of point *A* for the positions between these extremes, obtaining at least six points.

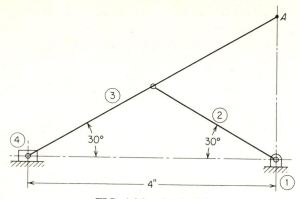

FIG. 4-26.   Prob. 4-11.

4-12.   (*a*) Draw the mechanism (Fig. 4-27, Tchebicheff's) in the position shown. (*b*) Draw the phase of the mechanism where link 2 is rotated to its extreme counterclockwise position. (*c*) Draw the phase of the mechanism where link 4 is rotated to its extreme clockwise position. (*d*) Draw the path of the midpoint of link 3 for the positions between these extremes, obtaining at least eight points.

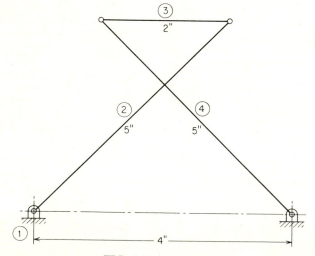

FIG. 4-27.   Prob. 4-12.

4-13. (*a*) Draw the mechanism (Fig. 4-28) in the position shown. (*b*) Determine the path of point *A* on link 2 by establishing 12 points corresponding to every 30° of crank rotation.

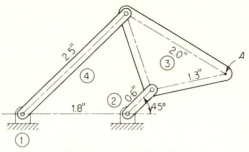

FIG. 4-28. Prob. 4-13.

4-14. First make a kinematic drawing of the film-transport mechanism (Fig. 4-29) in the position shown, transferring distances directly from the book (either full size or double size). Next, draw the path of point *B* by obtaining at least 10 points. *Suggestion:* Make small translucent overlays (skeleton outlines) of links 3 and 5 with small holes at points *A* and *B*, respectively. Then the locus of point *A* can be determined by moving the overlay of link 3 to various positions, keeping points *C* and *D* on their constrained arcs and marking the corresponding position for point *A* for each position. Then follow a similar procedure to obtain the locus of point *B*.

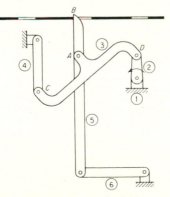

FIG. 4-29. Prob. 4-14.

centros (instantaneous centers)

5-1. General

For mechanisms involving levers, rollers, gears, pulleys, and cams, the motions are fairly easy to trace because the various links are always rotating about fixed centers or translating along fixed paths. Because of the relationship between the angular velocity of a rotating body and the linear velocity of any point on the body, the analysis of fixed-center links is simplified.

For mechanisms involving floating links (also called connecting rods or couplers) like link 3 in Fig. 5-1, however, special methods of analysis are required, involving such concepts as relative motion, instantaneous motion, and centros.

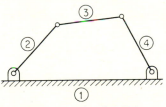

FIG. 5-1. Floating link.

5-2. Absolute and Relative Motion

Regardless of which link in a constrained kinematic chain is held stationary, the motion of any link relative to any other is always the same and always predictable. The term *stationary link* is merely a convenient term, since nothing in the universe is actually stationary. Therefore, the motion of a particular link must always be described relative to some other link. The term *absolute motion* means that the motion is described in terms of the reference link or frame, which is assumed for purposes of analysis to be at rest. The term *relative motion* means that the motion is described in terms of another link that is itself in motion with respect to the frame.

5-3.   Instantaneous Motion

The use of centros is based on the concept of instantaneous motion, and regardless of how complex the plane motion of the body may be, its instantaneous motion may be considered to be pure rotation.

In Fig. 5-2, if link 2 is displaced as shown, lines AA' and BB' represent the displacements of the ends of the link. If perpendicular bisectors are drawn for these two displacements and if their intersection at O is found, it is evident that $OA = OA'$ and $OB = OB'$ and that this displacement of link 2 *could* have been accomplished by a rotation about O. This may be easier to see if link 2 is visualized as one side of the triangle ABO that has rotated about its vertex O to the position $A'B'O$. If the distances AA' and BB' are considered infinitely small, then point O can be considered the instantaneous center (or centro) for this motion.

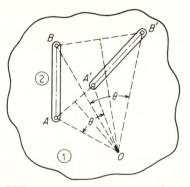

FIG. 5-2.   Instantaneous motion considered pure rotation.

5-4. Centros

A centro is the instantaneous center of rotation of one body with respect to another. In Fig. 5-2, point O is the centro for the instantaneous motion of link 2 relative to link 1. A centro may be thought of as:

1. A point common to two bodies and, therefore, having no relative motion between these bodies (either momentarily or permanently)
2. A point common to two bodies and having the same instantaneous linear velocity in each (either momentarily or permanently)
3. A point on one body about which another body is rotating (either momentarily or permanently)

5-5. Centro of a Body Where the Directions of Two Points Are Known

If the directions of any two points on a moving planar body are known, its centro can be found. In Fig. 5-3*a*, if points A and B in

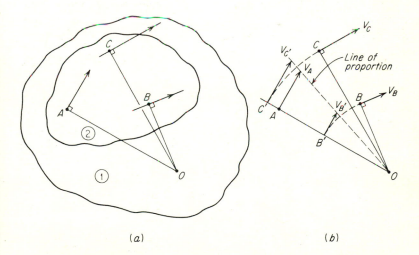

(*a*)   (*b*)

FIG. 5-3.   Locating centro of body knowing directions of two points.

link 2 have instantaneous motions with respect to link 1 in the directions indicated, their centers of rotation must be somewhere along lines perpendicular to these directions, since points on a rotating body always move in a direction perpendicular to their radii. Therefore, the intersection of the two radii locates the centro at O, which, for the instant shown, is the center of rotation of link 2 with respect to link 1. Now that the centro is located, the instantaneous direction of any other point on link 2 can be determined. Furthermore, if the instantaneous linear velocity of one of these points were known, the instantaneous velocities of the other points could be determined, because the linear velocities of points on a rotating body are proportional to their radii [Eq. (3-5)]. For example, if the velocity of point A were known, the vector V_A could be drawn to some scale, and a *line of proportion* could be drawn from its terminus (arrowhead) to the center of rotation, as shown in Fig. 5-3b. The other two points, B and C, could be rotated so that their radii coincided with that of A. The lengths of vectors V_B and V_C would then be determined by the line of proportion. These vectors could then be counterrotated to their original positions. This method of determining velocities is discussed in more detail in Chap. 6 in the section on the centro method.

5-6.   Centro Notation

The centro for any two links moving relative to each other is designated conventionally by the two numbers or letters used to identify the links. For example, if the two links were designated as *1* and *2*, their centro would be labeled 12 (referred to as *one two*, not twelve). If the links were designated as A and B, the centro would be *ab*. It is also conventional to show the two numbers or letters in numerical or alphabetical order.

5-7.   Primary Centros

For a given mechanism in a particular phase (position), there exists a centro for every pair of links. The *primary centros* are the centros that can be located by inspection, such as 12, 23, 34, and 14 in Fig. 5-4. The locations of centros 13 and 24 are not so obvious and must be located through the application of Kennedy's theorem, which is discussed in Art. 5-11.

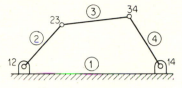

FIG. 5-4.   Primary centros.

5-8.   Centros of Pin-jointed Links

When two links in a mechanism are connected by a pin joint, it is obvious that the centro for all possible positions of these two links is at the pivot point, as shown in Fig. 5-5. Even if the pin joint moves as 23 does in Fig. 5-4, it is obvious that the pivot can always be considered a point common to both links and must have the same instantaneous linear velocity regardless of which link it is considered to belong to.

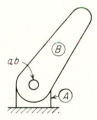

FIG. 5-5.   Centro at pin joint.

5-9.   Centros of Rolling Links

RULE

The centro for two links in rolling contact (no slipping) is located at their point of contact.

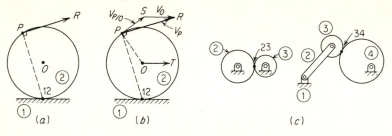

FIG. 5-6.   Centros of rolling links.

If wheel 2 rolls along link 1 without slipping, as shown in Fig. 5-6*a*, the centro 12 is at the point where 1 and 2 make contact. This is the only point that can be considered as being on 1 and 2, and it has the same instantaneous linear velocity (zero in this case) in either. The rolling of wheel 2 can be thought of as a series of minute rotations about each "element" of the wheel as it contacts link 1. If point *P* has a velocity *PR* relative to link 1 and if link 1 is stationary, then *PR* is the *absolute* velocity of *P*.

It is interesting to note that, in Fig. 5-6*b*, point *P* has a velocity *PS* relative to the center of the wheel *O* and that the center *O* has a velocity *OT* relative to link 1. The absolute velocity V_P of *P* is equal to the vector sum of the velocity V_O of *O* and the velocity $V_{P/O}$ of *P* relative to *O*, which can be expressed as $V_P = V_O +\!\!\!\!+ V_{P/O}$. Figure 5-6*c* shows two other examples of centros of rolling links.

5-10.   Centros of Sliding Links

RULE

The centro for two links having sliding contact must lie somewhere along their common normal (their common normal being perpendicular to their common tangent).

STRAIGHT SLIDER

It is evident in Fig. 5-7*a* that the center of rotation of the slider is at infinity (a straight line is a portion of a circle having an infinitely

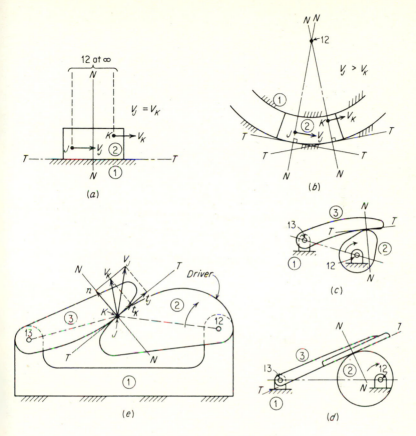

FIG. 5-7. Centros of sliding links.

large radius). Every point on a straight slider has the same velocity—both direction and magnitude—and so lines drawn perpendicular to the velocities of various points would be parallel to each other (and parallel to the common normal of the surfaces), indicating that the center of rotation is at infinity. The centro 12, in this case, could be considered to be on *any* line parallel to the common normal line, since parallel lines meet at infinity.

CURVED SLIDER

It is evident in Fig. 5-7*b* that the linear velocity of any point on a curved slider must be in a direction tangent to the curved path of the slider. Therefore, perpendiculars drawn from two or more points (not on the same radial line) will intersect, locating the center of rotation of the slider. Thus, the centro for the curved slider must lie on all these common normals. It is important to note that, *since sliders make area contact, they have an infinite number of common normal lines.*

SLIDING CONTACT: GENERAL

In the general case of two curved surfaces (Fig. 5-7*c*) or one curved and one flat surface (Fig. 5-7*d*) making contact, there is only one-point contact (within the plane of the drawing) and, therefore, only one common normal line. At this stage, then, it is not possible to locate their common centro definitely. It is only possible to say that the centro must lie somewhere along their common normal.

PROOF

In Fig. 5-7*e*, link 2 is the driver and link 3 is the follower. Points J and K are the points of contact on each link. The velocity of point J is represented by the vector V_J, and its direction is perpendicular to its radius J-12. The velocity of point K is represented by the vector V_K and its direction is perpendicular to its radius K-13. Since the velocities of these two points are different, point JK cannot be the centro.

If the two velocities V_J and V_K are each broken into two components, one along the common normal NN and one along the common tangent TT, it is evident that the two normal components must be equal; otherwise, links 2 and 3 would separate or crush each other. Therefore, the only relative motion that can take place between points J and K is along the common tangent TT, and this motion is represented by the difference in the tangent vectors t_J and t_K. Relative motion along the common tangent could only be achieved by rotation.about a center somewhere along the common normal line NN. *Therefore, the centro 23 must lie somewhere along NN, the common normal.*

To pinpoint the location of the centro along the common normal line, it is necessary to consider Kennedy's theorem.

5-11. Kennedy's Theorem

Kennedy's theorem states that *any three bodies (links) having plane motion relative to one another have three centros, all of which lie on a straight line.*

To prove this theorem, let 1, 2, and 3 in Fig. 5-8 be three links moving relative to one another. The three centros involved are:

$$1 \text{ relative to } 2: 12$$
$$1 \text{ relative to } 3: 13$$
$$2 \text{ relative to } 3: 23$$

Centros 12 and 13 are primary centros and their locations are obvious. In considering the location of 23, it must be remembered that this centro is a point common to both 2 and 3, and that it must have the same linear velocity in both links.

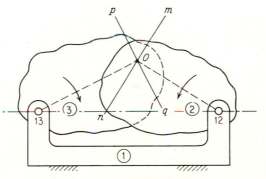

FIG. 5-8.   Kennedy's theorem.

Assume first, then, that centro 23 lies at point O. When O is regarded as a point on 2, it has a radius O-12 and moves along an arc about 12 in an instantaneous direction mn. If O is next considered as a point on 3, it has a radius O-13 and moves along an arc about 13 in an instantaneous direction pq. Therefore, since the directions *mn* and *pq* are not the same, point *O* could not possibly be common to both 2 and 3, and point *O* could not be the location of centro 23.

For a point to have the same direction or velocity in both 2 and 3, it would have to lie somewhere along the line of centers 12-13. Hence, *all three centros must lie along a straight line.* The use of Kennedy's theorem is indispensable in locating centros in a mechanism; its application will be illustrated in Art. 5-13.

Referring back to Fig. 5-7*c* to *e*, where it was shown that the centro for two sliding links must lie along their common normal, it is now obvious that the missing centro 23 in each of these cases must lie at the intersection of the common normal and the line of centers 12-13.

5-12.   Number of Centros in a Mechanism

The number of centros for n links can be expressed as the number of combinations of n things taken two at a time, which reduces to the expression

$$N = \frac{n(n-1)}{2} \tag{5-1}$$

EXPLANATION

The basic expression for combinations of n things taken r at a time is[1]

$$C^n_r = \frac{n!}{r!\,(n-r)!}$$

For n things taken *two* at a time the expression is

$$C^n_2 = \frac{n!}{2!\,(n-2)!} = \frac{n!}{2(n-2)!}$$

Showing this expression expanded into the factors indicated by the factorials gives the expression

$$C^n_2 = \frac{n(n-1)(n-2)(n-3)(n-4)\cdots(1)}{2(n-2)(n-3)(n-4)\cdots(1)}$$

Since every factor in the numerator to the right of $(n-1)$ cancels with

---

[1] The development of this expression can be found in most college algebra or probability textbooks.

every factor in the denominator to the right of 2, the expression reduces to

$$N = \frac{n(n-1)}{2}$$

It should be pointed out that in practical work it is rarely necessary to locate all centros in a mechanism; only those which are essential to the analysis should·be located.

5-13.   Table of Centros

A table of centros provides a systematic way of listing all the possible centros of a mechanism and of showing which centros have been located. As shown in Fig. 5-9*a* the numbers (or letters) of all the links are listed across the top of the table in order. In the first column are listed all the centros that contain the number (or letter) at the top of the column combined with those to the right. In the second column the number at the top of that column is combined with those to the right, and so on.

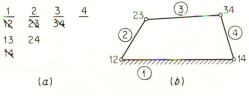

FIG. 5-9.   Table of centros.

A line is drawn through each centro in the table that can be located by inspection—the primary centros. ·For the mechanism shown in Fig. 5-9*b*, the locations of centros 12, 23, 34, and 14 are known, so they are *lined out* in the table. This makes it clear that centros 13 and 24 are those yet to be found. To apply Kennedy's theorem in locating these centros, it is convenient to draw smaller supplementary tables. The use of this method is illustrated in the following example.

EXAMPLE 5-1. USING TABLES OF CENTROS

In Fig. 5-10a we are required to find all centros for the four-link mechanism. The four primary centros are located first and labeled as shown.

TO FIND CENTRO 13:

If links 1 and 3 are thought of in combination with any third link,
according to Kennedy, the three centros for these links must lie on a
straight line. Furthermore, the straight line is defined if two of the three
centros are already located.

1. Consider 1 and 3 in combination with 2, and make a supplementary
    table of centros.

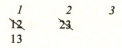

Since 12 and 23 are already located, a straight line through these
two centros must also contain 13. This line is shown in Fig. 5-10b.
Another line is needed to establish definitely the location of 13.

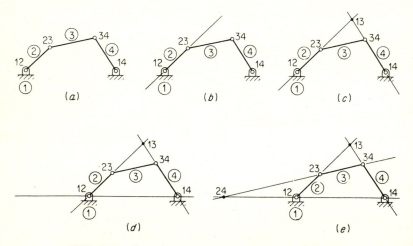

FIG. 5-10. Locating centros in a four-link mechanism.

2. Next consider 1 and 3 in combination with 4, and make a second supplementary table of centros.

1   *3*   *4*

13   34

14

Since 14 and 34 are already located, a straight line through these two centros must also contain 13. This line is shown in Fig. 5-10*c*. The location of 13 is now established by the intersection of these two lines.

TO FIND CENTRO 24:

To find 24 it is necessary to use links 2 and 4 in combination with first one link and then with a second so as to obtain two intersecting lines, as for 13.

1. Consider links 2 and 4 in combination with 1. The known centros 12 and 14 establish the line shown in Fig. 5-10*d*.

1   *2*   *4*

12   24

14

2. Next consider links 2 and 4 in combination with 3. The known centros 23 and 34 establish the line shown in Fig. 5-10*e*.

2   *3*   *4*

23   34

24

The intersection of these two lines locates 24.

5-14. Centro Diagram (Circle Diagram)

A centro diagram provides an effective graphic method of keeping a running account of which centros have been found and which need to be found. It further indicates which combinations of centros may be used in applying Kennedy's theorem.

To construct a centro diagram, a small circle is drawn, 1 to 2 in. in diameter. The circumference of the circle is then divided into as many equal spaces as there are links in the mechanism, and the resulting points are identified with numbers (or letters) corresponding to the links. For the four-link mechanism shown in Fig. 5-9*b*, it appears at first, as shown in Fig. 5-11*a*. For any two points (representing links) on the diagram, there is a connecting straight line (representing their common centro). First, all lines representing centros whose locations are known are drawn in heavy as shown in Fig. 5-11*b*. The centros remaining to be located may then be represented by dashed lines as shown in Fig. 5-11*c*.

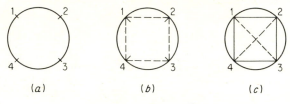

FIG. 5-11.   Centro diagram.

Note that the lines of the diagram form triangles. Each of these triangles has the numbers representing three links at its vertices, and its three sides can be thought of as representing the three centros for these three links. From Kennedy's theorem, then, the three centros represented by the three sides of a triangle must lie on a straight line. Then if two sides of a triangle are drawn in heavy, indicating that two

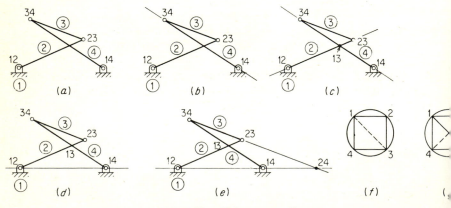

FIG. 5-12.   Using centro diagram.

of the three centros are located, a line can be drawn that contains the third centro. To locate a particular centro, it is necessary to establish two such lines. Therefore, to locate a centro, *two* triangles must be found in the diagram that have two known sides *and* have as their unknown side the centro being sought.

EXAMPLE 5-2.   USING CENTRO DIAGRAM

It is required to find centros 13 and 24 in the mechanism shown in Fig. 5-12*a*.

TO LOCATE CENTRO 13:

It should be observed in the centro diagram (Fig. 5-12*f*), that the dashed line representing unknown centro 13 is common to triangles 143 and 123, both of whose other two sides represent centros whose locations are known. According to triangle 143, centros 13, 14, and 34 must lie on a straight line, and, since 14 and 34 are already located, they establish line 14-34 which contains 13. This line is indicated in Fig. 5-12*b*. According to triangle 123, centros 12, 13, and 23 must lie on a straight line, and, since 12 and 23 are already located, they establish a second line that contains 13, as shown in Fig. 5-12*c*. The intersection of these two lines locates centro 13.

TO LOCATE CENTRO 24:

Two triangles that contain 24 and have two known sides are 214 and 234, as shown in Fig. 5-12*g*. Triangle 214 contains the known centros 12 and 14, thus establishing one line containing 24, as shown in Fig. 5-12*d*. Triangle 234 contains the known centros 23 and 34, thus establishing a second line containing 24, as shown in Fig. 5-12*e*. The intersection of these two lines locates centro 24.

EXAMPLE 5-3.   SLIDER CRANK

It is required to find centros 12 and 34 in the slider-crank mechanism shown in Fig. 5-13*a*. The links were deliberately numbered nonconsecutively in this example to alter the appearance of the centro diagram.

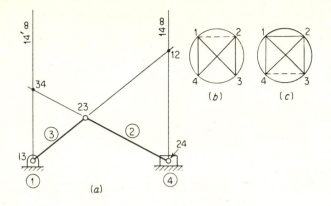

FIG. 5-13. Slider crank.

TO LOCATE CENTRO 12:

Use triangles 142 and 132 in the centro diagram of Fig. 5-13*b*. Triangle 142 is made up of centros 12, 14, and 24, which must lie on a straight line. Centros 14 and 24 are already located, so they establish one of the two required lines.

Triangle 132 is made up of centros 12, 13, and 23, which must lie on a straight line. Centros 13 and 23 are already located, so they establish the other required line.

Centro 12 is thus located at the intersection of lines 14-24 and 13-23 (extended), as shown in Fig. 5-13*a*.

TO LOCATE CENTRO 34:

Use triangles 324 and 314 in the centro diagram in Fig. 5-13*c*. Triangle 324 is made up of centros 23, 24, and 34, which must lie on a straight line. Centros 23 and 24 are already located, so they establish one of the required lines.

Triangle 314 is made up of centros 13, 14, and 34, which must lie on a straight line. Centros 13 and 14 are already located, so they establish the other required line. *It is important to note that centro 14 is at infinity and that, since parallel lines meet at infinity, 14 may be shown on any line parallel to the common normal of the slider.* In this case, to draw a line through centros 13 and 14, it is necessary to draw a line through 13 and parallel to the line labeled 14 ∞. This new line is labeled 14′ ∞. The intersection of the two lines 23–24 and 13–14′ ∞ establishes the location of centro 34.

5-15.   Finding Centros in Complex Mechanisms

Finding centros for mechanisms having more than four links becomes much more confusing, and the chances for making mistakes increase considerably. It occasionally happens that two triangles in the centro diagram will indicate two lines that are unsatisfactory for one of the following reasons: (1) the two lines are coincident, thereby producing no intersection; (2) the two lines are so nearly parallel that their intersection is indefinite; (3) one of the lines is established by two centros that are too close together to establish accurately the direction of the line; or (4) the two known centros in a triangle coincide, thereby establishing no line. When one of these situations occurs, another triangle must be found. If a third triangle cannot be found, the location of another centro must be attempted. As more centros are found, it becomes increasingly easy to find triangles to use.

Figure 5-14a shows a six-link mechanism with all of its primary centros indicated on the mechanism as well as in the centro diagram (Fig. 5-14b). Eight of the fifteen centros remain to be located.

Figure 5-14d to k shows the centro diagram as each new centro is found, and the triangles involved are shown shaded. For example, to find centro 13, Fig. 5-14d shows that triangles 123 and 143 were used. These triangles produced lines 12-23 and 14-34, whose intersection located centro 13.

Figure 5-15 shows how the procedure may be tabulated to record the sequence of the solution. Since in the later stages of the solution there will usually be more than two triangles that could be used to locate a particular centro, there exist opportunities to confirm some of the locations. The tabulation shows that centros 25, 46, and 45 were confirmed. The extra intersecting line representing each confirmation is shown as a dashed line in Fig. 5-14c. It should be noticed that the first attempt to confirm centro 45 resulted in line 34-35 which coincides with line 14-15, so a fourth triangle 465 was used.

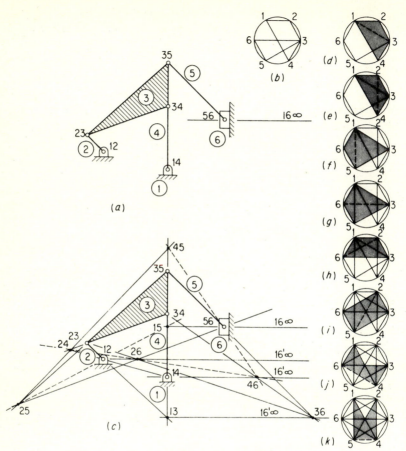

FIG. 5-14.   Locating centros in a six-link mechanism.

5-16. Alternative Method for Finding Centro of Floating Link

In Examples 5-1 to 5-3, the centros for the floating links (Fig. 1-4) could be obtained without the use of Kennedy's theorem. In each case, the directions of travel of both ends of these links are known,

To Locate		Triangles used	Intersecting lines	Confirmed with	
				△	line
*(d)	13	123 ∉ 143	12-23 ∉ 14-34		
(e)	24	234 ∉ 214	23-34 ∉ 21-14		
(f)	15	135 ∉ 165	13-35 ∉ 16-65		
(g)	36	316 ∉ 356	13-16 ∉ 35-56		
(h)	26	216 ∉ 236	21-16 ∉ 23-36		
(i)	25	235 ∉ 265	23-35 ∉ 26-56	215	12-15
(j)	46	416 ∉ 436	14-16 ∉ 34-36	426	24-26
(k)	45	415 ∉ 425	14-15 ∉ 24-25	~~435~~	~~34-35~~
				465	46-56

*Letters refer to appropriate portions of fig. 5-14.

FIG. 5-15. Tabulation of centros.

so it is necessary only to draw lines perpendicular to these directions of travel, as shown in Fig. 5-16. In each case, the centro obtained is the centro for the floating link with respect to the frame, i.e., centro 13. This method was explained in Art. 5-5.

5-17. Zero-velocity Centros

It should be noted that any centro containing the number (or letter) of the frame (fixed link) represents a point on a moving link that has zero velocity with respect to the frame for the particular phase shown. It cannot have a velocity because, by definition, it can be considered a point on either link and must have the same linear velocity in both. If it can be considered a point on the frame, its velocity must be zero.

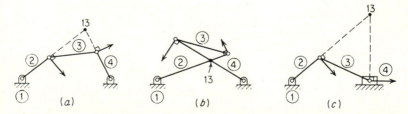

FIG. 5-16. Alternative method for finding centro of floating link.

5-18.   Centrodes

The *path* of a centro of a moving link is called a *centrode*. When one link has constrained motion relative to another, their centrodes always make contact at a point. As motion continues, the two centrodes roll on each other. It is possible, therefore, to substitute for a given mechanism an *equivalent mechanism* consisting of two rolling surfaces. While quite interesting, no use is made of this concept in this text.[1]

5-19.   Labeling Centros That Lie off the Paper

Quite often one or more of the centros for a particular phase of a mechanism lie beyond the edge of the drawing paper. It is important to indicate the location of such centros in order to avoid confusion in locating other centros and to make it clear which centros are not available for velocity constructions (which are discussed in the next chapter). Figure 5-17 illustrates how such centros should be labeled.

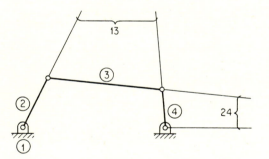

FIG. 5-17.   Locating centros that lie off the paper.

Problems

5-1 to 5-14.   Locate and label all centros for Figs. 5-18 to 5-31.

[1] For a good treatment of this topic, see Joseph E. Shigley, *Kinematic Analysis of Mechanisms*, McGraw-Hill Book Company, New York, 1969.

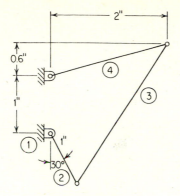

FIG. 5-18.   Prob. 5-1.

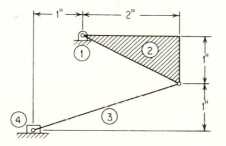

FIG. 5-19.   Prob. 5-2.

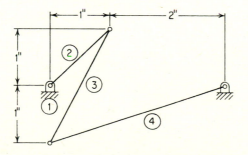

FIG. 5-20.   Prob. 5-3.

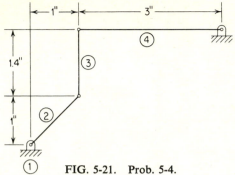

FIG. 5-21.   Prob. 5-4.

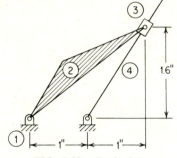

FIG. 5-22.   Prob. 5-5.

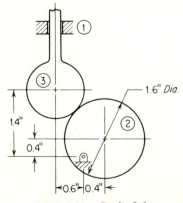

FIG. 5-23.   Prob. 5-6.

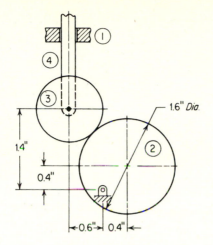

FIG. 5-24. Prob. 5-7.

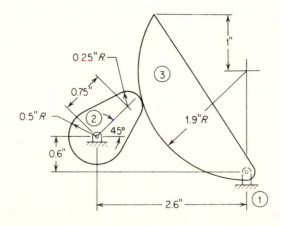

FIG. 5-25. Prob. 5-8.

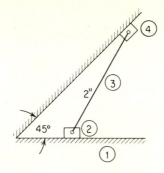

FIG. 5-26. Prob. 5-9.

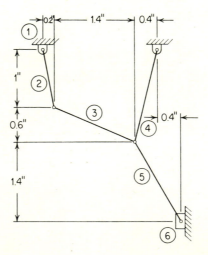

FIG. 5-27. Prob. 5-10.

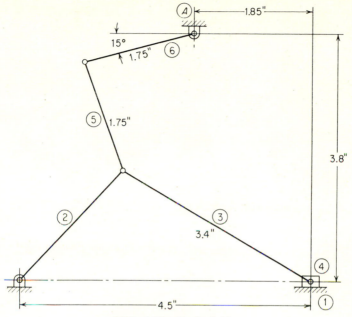

FIG. 5-28. Prob. 5-11.

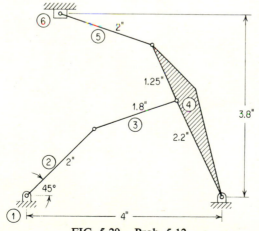

FIG. 5-29. Prob. 5-12.

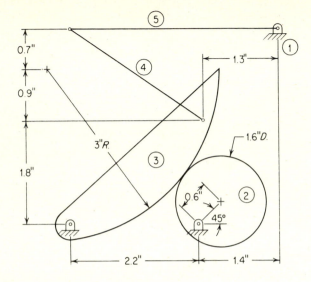

FIG. 5-30. Prob. 5-13.

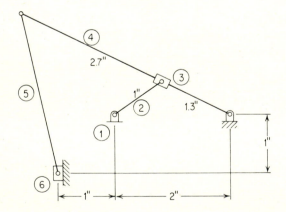

FIG. 5-31. Prob. 5-14.

velocities in mechanisms

6-1. Introduction

This chapter presents the four most important graphical methods for finding velocities in mechanisms:

1. The centro method
2. The parallel-line method
3. The component method
4. The relative-velocity method

All four methods are presented because each has distinct advantages for particular situations. While the relative-velocity method is by far the most important because of its value in connection with acceleration analysis, the other methods often afford a better insight into the relationships existing between the various velocities.

6-2. Relationship of Velocities of Points on a Rotating Body

A basic construction that is used in connection with all four of the velocity methods is based on the fact that the linear velocities of two points on a rotating body are proportional to their distances from the center of rotation [Eq. (3-5)].

In Fig. 6-1, body A containing points X and Y is rotating about point O. If the linear velocity V_Y of point Y is known, a right triangle can be drawn using the velocity vector and its radius as the two perpendicular sides. The hypotenuse, extending from the terminus of the vector to the center O, is called the *line of proportion*. Point X can then be rotated to a point where its radius coincides with that of Y. The line of proportion then determines the magnitude of the velocity V_X of X. This vector can then be counterrotated to its original position.

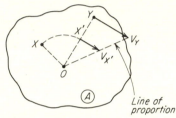

FIG. 6-1. Linear velocities of points on a rotating body.

6-3.   The Centro Method: General Approach

The centro method is based on the following facts: (a) a centro common to two links is a point that can be considered on either link and has the same velocity in each, and (b) the linear velocity of a point on a rotating body is proportional to its radius. Figure 6-2 shows a

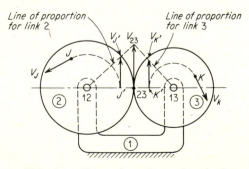

FIG. 6-2.   Illustration of centro method.

three-link mechanism consisting of two rollers and a frame. These three links have three centros 12, 13, and 23, which lie on a straight line, as shown. If the velocity of J is known, then it is possible to find the velocity of any other point on link 2 by the construction shown in the previous article. Centro 23 can be considered a point on link 2 and its velocity found by revolving the known velocity V_J to the line of centers, where it serves to establish the line of proportion for link 2. This line of proportion is extended to establish the velocity of the centro 23. Now that V_{23} is known, 23 can be considered a point on link 3 and can be used to determine the line of proportion for this link. Then point K can be rotated to the line of centers and the magnitude of its velocity determined, after which it may be counterrotated to its original position.

From this illustration, the following facts should be noted:

1. That three links are involved: the frame, the link containing the known velocity, and the link containing the unknown velocity.
2. That three centros are involved: the *common centro* for the two moving links and the centros for each of the moving links with respect to the frame. The last two are referred to as *zero-velocity centros.*
3. That the two *zero-velocity centros* are the vertices of the two triangles formed along the line of centers and are the points about which the rotations take place.

The above illustration and observations can be expanded and restated as a *general method* for finding velocities in more complex mechanisms. The basic problem usually involves knowing the velocity of a particular point in a mechanism and being interested in determining the velocity of some other point in the mechanism.

The steps to be followed are:

Step 1.   Determine which three links are involved: the link containing the point of known velocity, the link containing the point for which the velocity is desired, and the frame.

Step 2.   Determine the *line of centers* by making up a table of centros for the three links involved, and label this line *LC*. For example, if links 1, 3, and 6 are involved, the line of centers would be established by the three centros 13, 16, and 36.

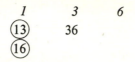

Step 3.   Determine which two of these three centros are the zero-velocity centros, and circle these on the drawing and in the table of centros. The zero-velocity centros are those containing the number of the frame (the frame is assumed to be 1 in this case), and they are the centers of rotation for the two moving links.

Step 4.   Rotate the known velocity to the line of centers. It must be rotated about the center of rotation of the link that it is on.

Step 5.   Draw the line of proportion for the link containing the known velocity by drawing a line from its center of rotation to the terminus of the known velocity in its rotated position.

Step 6.   Determine the velocity of the common centro using the line of proportion just established.

Step 7.   Draw the line of proportion for the link containing the unknown velocity by now considering the common centro as a point on this link. This is done by drawing a line from the terminus of the velocity of the common centro to the center of rotation for this link.

Step 8.   Rotate the point for which the velocity is desired to the line of centers and determine the magnitude of its velocity by the line of proportion just established.

Step 9.   Counterrotate the vector to the point's original position.

EXAMPLE 6-1.   CENTRO METHOD: POINTS ON ADJACENT LINKS

In Fig. 6-3*a*, the velocity of point *J* is known, and the velocity of point *K* is required.

SOLUTION

1.   The links involved are:
     1, frame
     2, link containing the known velocity
     3, link containing the unknown velocity

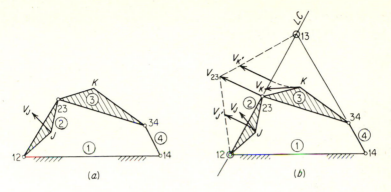

FIG. 6-3.   Centro method: points on adjacent links.

2. The line of centers is determined by centros 12, 13, and 23, and is drawn in and labeled LC (Fig. 6-3b).

	1	*2*	*3*
Zero-velocity centros	(12) (13)	23	← Common centro

3. Centros 12 and 13 are the zero-velocity centros (they are circled); centro 23 is the common centro.
4. V_J is rotated to the line of centers, where it is labeled $V_{J'}$. Note that V_J is rotated about centro 12, which is the center of rotation of link 2.
5. The line of proportion for link 2 is drawn from centro 12 through the terminus of $V_{J'}$ and is labeled LP_2.
6. The velocity of the common centro V_{23} is then established and its vector drawn.
7. The line of proportion for link 3 is drawn from the terminus of V_{23} to centro 13 and is labeled LP_3.
8. Point K is rotated to the line of centers, where its magnitude is determined by the line of proportion LP_3 and is labeled $V_{K'}$. Note that point K is rotated about centro 13, which is the center of rotation of link 3.
9. The vector $V_{K'}$ is counterrotated to its true position and labeled V_K.

EXAMPLE 6-2.   CENTRO METHOD: POINTS ON OPPOSITE LINKS

In Fig. 6-4*a*, the velocity of point *J* is known, and the velocity of point *K* is required.

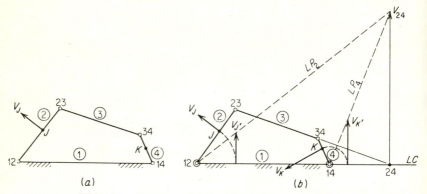

(*a*)   (*b*)

FIG. 6-4.   Centro method: points on opposite links.

SOLUTION

1. The links involved are:
    1, frame
    2, link containing known velocity
    3, link containing unknown velocity
2. The line of centers is determined by the following centros:

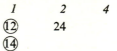

3. Centros 12 and 24 are the zero-velocity centros (they are circled); centro 24 is the common centro.
4. V_J is rotated to the line of centers and is labeled $V_{J'}$.
5. The line of proportion for link 2 is drawn from centro 12 through the terminus of $V_{J'}$ and is labeled LP_2.
6. The velocity of the common centro V_{24} is established by extending LP_2.
7. The line of proportion for link 4 is drawn from the terminus of V_{24} to centro 14 and is labeled LP_4.

8. Point K is rotated to the line of centers, where its magnitude is determined by LP_4 and is labeled $V_{K'}$. (*Note:* In this case point K was rotated clockwise to the line of centers; however, a counterclockwise rotation would have produced the same result.)
9. The vector $V_{K'}$ is counterrotated to its true position and labeled V_K.

EXAMPLE 6-3.   CENTRO METHOD: SLIDER CRANK

In Fig. 6-5, the velocity of point J is known, and the velocity of point L on the slider is required.

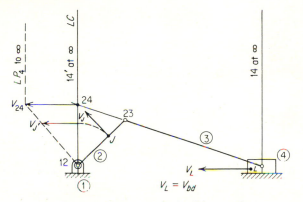

FIG. 6-5.   Centro method: slider crank.

SOLUTION

1. The links involved are:
   1, frame
   2, link containing known velocity
   3, link containing unknown velocity
2. The line of centers is determined by the following centros:

$$\begin{array}{ccc} 1 & 2 & 4 \\ \textcircled{12} & 24 & \\ \textcircled{14} & & \end{array}$$

3. Centros 12 and 14 are the zero-velocity centros (they are circled); centro 24 is the common centro.
4. V_J is rotated to the line of centers.

5. The line of proportion for link 2 is drawn from centro 12 through the terminus of $V_{J'}$ and is labeled LP_2.
6. The velocity of the common centro V_{24} is established by extending LP_2.
7. Now, if the common centro 24 is considered to be a point on slider 4, it is evident that $V_L = V_{24}$ because all points on a slider have the same velocity. In fact, if LP_4 is drawn from the terminus of V_{24} to the centro 14, which is at infinity, it turns out to be parallel to the line of centers as shown.

EXAMPLE 6-4.   CENTRO METHOD: SLIDING LINKS

In Fig. 6-6, the velocity of point E on link 2 is known, and the velocity of point K on link 3 is required. The point of contact is represented by point E on link 2 and point F on link 3.

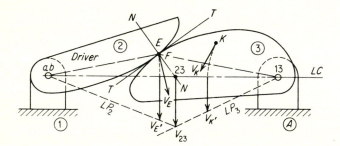

FIG. 6-6.   Centro method: sliding links.

SOLUTION

1. The known velocity V_E is rotated to the line of centers 12-23-13.
2. The line of proportion LP_2 is drawn through centro 12 and the terminus of $V_{E'}$ and extended to determine V_{23}, which is considered on link 2.
3. V_{23} is now considered on link 3 and is used to determine LP_3.
4. Point K is rotated to the line of centers, where its magnitude $V_{K'}$ is determined by LP_3.
5. The vector $V_{K'}$ is counterrotated to its true position V_K.

EXAMPLE 6-5. CENTRO METHOD: ROLLING LINK

In the case of the roller mechanism shown in Fig. 6-7, the velocity of the end of link 2 is known (centro 23) and the velocity of the contact point of link 4 (centro 34) is required. Since both points can be considered as points on link 3 and therefore instantaneously rotating about centro 13, the line of proportion for link 3 is drawn from centro 13 through the terminus of the known velocity V_{23}. Point 34 is then rotated about 13 until its radius coincides with that of 23, and the magnitude of its velocity is determined by the line of proportion.

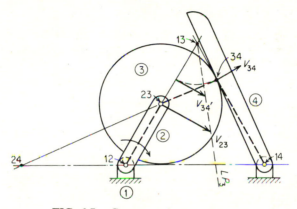

FIG. 6-7. Centro method: rolling links.

In a practical problem of this sort, the given data would probably be in the form of the angular velocity of link 2, and the angular velocity of link 4 would be required. In this case, the linear velocity of point 23 would be calculated using the relationship $V_{23} = r_{23}\omega_2$. Then the velocity of point 34 would be obtained in the manner of the example, after which the angular velocity of link 4 would be obtained using the relationship $\omega_4 = V_{34}/r_{34}$.

One further remark concerning this example. It should be noted that, for purposes of a velocity analysis, the roller mechanism could be replaced by the "equivalent" pin-jointed mechanism indicated by the dashed lines. The equivalency of this mechanism applies only to the particular phase shown, and it applies only to velocity analysis and not to acceleration analysis.

Example 6-6.   Centro Method: Six-link Mechanism

In the six-link mechanism shown in Fig. 6-8, the velocity of point A is known, and the velocity of point C is required. If the general approach is to be followed, the first step is to determine which three links are involved (one of the three must always be the frame). Since point A is on both links 2 and 3 and point C is on both links 5 and 6, it is apparent that there is more than one possible combination. The most obvious combination of links would be 1, 2, and 6, which would involve centros 12, 16, and 26. The other possible combinations are 1-2-5, 1-2-6, 1-3-5, and 1-3-6. Before deciding on which combination to use, the centro diagram should be prepared for the mechanism and all known centros indicated with solid lines. A specific combination should then be selected tentatively. For the particular combination selected, the centros required should be shown in the centro diagram. If the required centros can be located easily, then proceed with this particular approach. On the other hand, if the necessary centros cannot be located easily, then another combination of links will probably provide an easier solution. For this mechanism it turns out that the two combinations resulting in the centros easiest to find are 1-3-5 and 1-3-6.

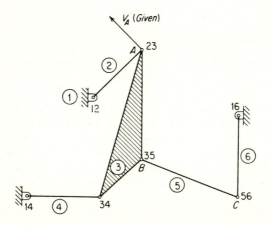

FIG. 6-8.   Centro method: six-link mechanism.

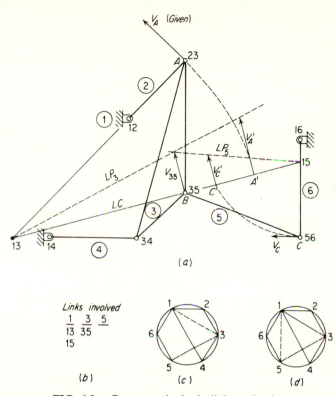

FIG. 6-9. Centro method: six-link mechanism.

Figure 6-9 shows the solution using links 1, 3, and 5. The table of centros (Fig. 6-9b) shows that centros 13, 15, and 35 are required to establish the line of centers. The centro diagram in Fig. 6-9c indicates that centro 13 can be located. The centro diagram in Fig. 6-9d indicates that centro 15 can be located as soon as 13 has been located.

The solution then consists of the following: establishing the line of centers through centros 13, 15, and 35; rotating V_A to the line of centers (about 13); establishing LP_3; using LP_3 to determine V_{35}, the velocity of the common centro; using V_{35} to establish LP_5; then determining V_C by rotating C to the line of centers (about 15) and utilizing LP_5.

Figure 6-10 shows the equally simple solution using links 1, 3, and 6. Note that centros 13 and 36 are needed. Figure 6-10c again indicates that centro 13 can be located and Fig. 6-10d indicates that centro 36 can be located as soon as 13 has been located.

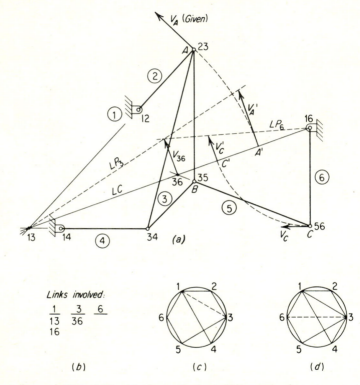

FIG. 6-10.   Centro method: six-link mechanism.

The pattern of the solution is the same as that described for Fig. 6-9. The solution differs only in that the line of centers is different—the common centro is now 36 rather than 35, and point *C* is rotated about 16 rather than 15.

Figure 6-11 shows the solution involving links 1, 2, and 5. The centro diagram in Fig. 6-11c indicates by dashed lines the centros to be found. It is more or less readily apparent (in the centro diagram) that

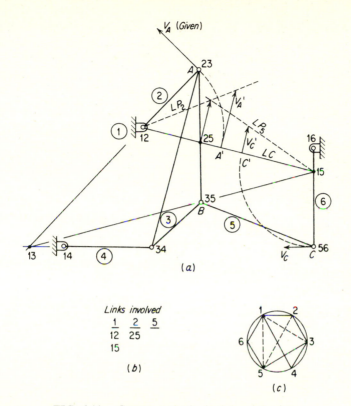

FIG. 6-11. Centro method: six-link mechanism.

once centro 13 is located, centro 15 can be located, and then 25 can be located. The resulting lines of centers and rotations are indicated in Fig. 6-11*a*.

Figure 6-12 shows the solution involving links 1, 2, and 6. Although this particular combination might intuitively be the first choice, it turns out to be the most difficult solution as far as locating centros is concerned. The centro diagram in Fig. 6-12*c* shows that only centro 26 must be located. A careful look at the situation, however, reveals the fact that, in order to locate 26, centro 36 must first be located, and, in order to locate 36, centro 13 must be located. This sequence is indicated by the centro

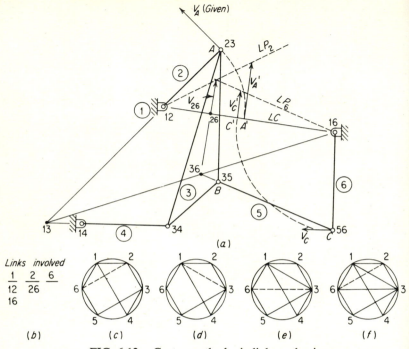

Links involved

1	2	6
12	26	
16		

FIG. 6-12. Centro method: six-link mechanism.

diagrams in Fig. 6-12d to f. Figure 6-12a shows the resulting solution, which is as straightforward as the other solutions once the necessary centros are located.

6-4. The Centro Method: Link-to-link Approach

The centro method is sometimes referred to as the *direct approach* since the general procedure permits going from any point in a mechanism to any other point regardless of how many links are in between, provided the necessary centros can be located and the line of centers established. The centro method is equally effective when used in a *link-to-link* fashion, however, and in many cases is the easier of the two approaches to follow.

In the *link-to-link* approach, each step consists of working with two points on the same link, one whose velocity is known and the other whose velocity is required. If the center of rotation of the link is known, a line of proportion can be quickly drawn from the terminus of the known velocity to the center of rotation; the point whose velocity is required is rotated so that its radius coincides with that of the known point, and the magnitude of its velocity is determined by the line of proportion. This basic construction was illustrated in Fig. 6-1, and was utilized in Fig. 6-7. Using this basic construction as many times as necessary, velocities may be "chased through" a mechanism in a link-to-link fashion.

EXAMPLE 6-7.   CENTRO METHOD: LINK-TO-LINK APPROACH
Figure 6-13 shows the link-to-link approach applied to the same mechanism that was involved in Example 6-6 (Figs. 6-8 to 6-12). With this approach, the first step involves working with points *A* and *B* which are both on link 3. A line of proportion for link 3 is established by drawing a line from the center of rotation of link 3 (centro 13) to the terminus of the known velocity V_A. Point *B* is then rotated about 13 until its radius

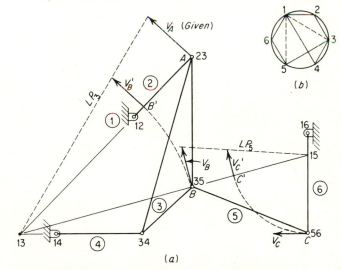

FIG. 6-13.   Centro method: link-to-link approach.

coincides with that of A, and the magnitude of its velocity is determined by LP_3, after which it is rotated back to its original position.

The second step involves working with points B and C, which are both on link 5. The center of rotation for link 5 (centro 15) must be located. Then a line of proportion for link 5 is established by drawing a line from centro 15 to the terminus of V_B. The magnitude of V_C is then found by rotating C so that its radius coincides with that of B and using LP_5 as shown.

6-5. The Parallel-line Method

The parallel-line method has an advantage over the centro method in some instances where the centro of a floating link is inaccessible (lies off the paper), and its construction is somewhat simpler. Its chief limitation is that it is comparatively indirect, for velocities must be transferred from link to link to get to the opposite side of a mechanism. In fact, the method is nothing more than a variation of the link-to-link approach of the centro method (Art. 6-4) whereby a parallel-line construction replaces the line of proportion in each step.

The parallel-line method, then, is restricted to situations involving two points on the same link, where the velocity of one point is known and it is desired to find the velocity of the other. Like the centro method, the parallel-line method requires the location of the centro of the link with respect to the frame. The steps involved are as follows (see Fig. 6-14):

1. Rotate the known velocity vector 90° such that it coincides with the radius of the point and points toward the center of rotation.
2. Draw a line through the terminus of this rotated vector and parallel to the line connecting the two points.

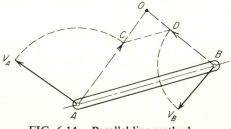

FIG. 6-14. Parallel-line method.

3. The point where this parallel line intercepts the radius of the point whose velocity is desired determines the magnitude of the velocity of this second point.

4. The segment thus obtained is rotated 90° in the opposite direction to that of the rotation of the given vector in step 2, and this results in the correct velocity vector for the second point.

PROOF

This method is based on (1) the geometry of similar triangles and (2) the proportionality of instantaneous linear velocities on a rotating link. In Fig. 6-14 it is obvious that

$$\frac{V_A}{OA} = \frac{V_B}{OB} \tag{1}$$

This can be rewritten as

$$\frac{V_A}{V_B} = \frac{OA}{OB} \tag{2}$$

Also, if lines AB and CD are drawn parallel, then by similar triangles

$$\frac{OA}{OB} = \frac{OC}{OD} \tag{3}$$

Then Eqs. (2) and (3) can be combined such that

$$\frac{V_A}{V_B} = \frac{OA - OC}{OB - OD} = \frac{CA}{DB}$$

Therefore, if CA is constructed equal to the magnitude of V_A, then DB will be equal to the magnitude of V_B.

Figure 6-15 compares the parallel-line method with the centro method. Note that the parallel-line method can be used even when the centro 13 is inaccessible.

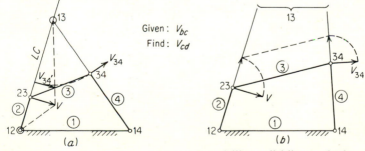

Given: V_{bc}
Find: V_{cd}

FIG. 6-15. Comparison of (*a*) centro and (*b*) parallel-line methods.

EXAMPLE 6-8. PARALLEL-LINE METHOD

In Fig. 6-16, the velocity of point A is known, and the velocities of D and E are required. It should be noted that although the velocities of points B and C are not required, they are needed as stepping stones to derive the required velocities.

SOLUTION

1. Figure 6-16a shows how the velocity V_B is found by rotating the known velocity V_A so that its radius coincides with that of B. Reverting to the centro method for obtaining velocities of points on pivoting links is much more effective than attempting to apply the parallel-line method. Figure 6-17 shows how it is possible to obtain the velocity of B by the parallel-line method.

2. Now that V_B is known, the velocity of point D can be found by utilizing the fact that points B and D are on the same link 3, whose center of rotation is centro 13. As shown in Fig. 6-16a, vector V_B is rotated toward the center of rotation. A line is then drawn through its terminus and parallel to line BD. The point where this line intercepts the radius of D determines the magnitude of V_D, and it merely has to be rotated to its correct position perpendicular to its radius.

3. As shown in Fig. 6-16b the velocity of C is obtained by starting again from the rotated V_B—only this time a line is drawn parallel to line BC. (It would also be possible to start from the rotated position of V_D and draw a line parallel to DC.)

4. The velocity of E is then obtained by the centro method, considering point C to be on link 4, drawing a line of proportion for link 4, and rotating E so that its radius coincides with that of C. Again it would be possible to obtain the velocity of E by the parallel-line method using a construction identical to that shown in Fig. 6-17.

TO FIND V_D WITHOUT USING CENTRO 13 (Fig. 6-16c):

If the centro 13 were not readily accessible and therefore the radius of D not available, then the intersection of the line parallel to line BD and the radius of D would not be available. In this case, merely draw the line parallel to BD to an indefinite length. Then draw a second line through the terminus of the rotated V_C vector and parallel to line CD.

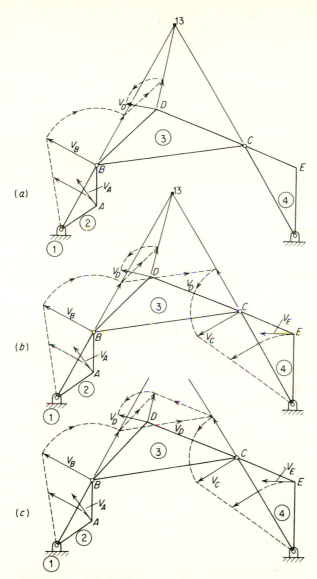

FIG. 6-16. Parallel-line method: four-link mechanism.

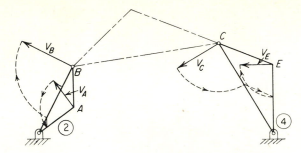

FIG. 6-17.   Parallel-line method: pivoted link.

The intersection of these two lines determines the terminus of the rotated velocity vector V_D. In effect, the velocity of D is found by converging on it from two other points on the same link.

EXAMPLE 6-9.   PARALLEL-LINE METHOD: SIX-LINK MECHANISM

In Fig. 6-18*a*, the velocity of A is known, and the velocity of C is required.

In this case, the velocity of B is found first since A and B are on the same link 3. Vector V_A is rotated to point toward the center of rotation of link 3. A line is drawn through the terminus of this rotated vector, parallel to line AB. The point where this line intercepts the radius of B determines the magnitude of V_B. The actual velocity of B must be perpendicular to link 4.

Now that the velocity of B is known, the velocity of C can be obtained by working with points B and C, both of which are on link 5. The center of rotation of link 5 must be located (centro 15), and the centro diagram in Fig. 6-18*b* indicates that centro 15 can be found easily. After locating centro 15, rotate the vector V_B toward 15, then draw a line through its terminus in this position and parallel to line BC. The point where this line intercepts the radius of C provides the magnitude of V_C. This vector is then rotated 90° to provide the correct velocity of C.

6-6.   The Component Method

The component method utilizes rectangular components of velocity vectors and is based on the fact that *two points on the same rigid link*

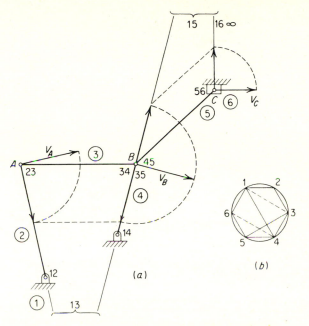

FIG. 6-18. Parallel-line method: six-link mechanism.

must have the same component of velocity in the direction of the line joining the two points. This common component is the orthogonal component of each velocity (see Art. 2-7) and is referred to as the *parallel component.*

In Fig. 6-19, if the velocity of point *A* is known and if the direction of *B* is known (perpendicular to its radius), it is possible to obtain the velocity of *B* as follows:

1. Resolve the known velocity into two rectangular components, one parallel to a line joining the two points and one perpendicular to the line.

2. The component of the known velocity that is parallel to the line joining the two points must be equal to a component of the velocity of the other point along this line (otherwise the link would crush or pull apart).

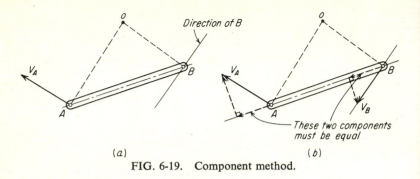

FIG. 6-19.   Component method.

3. The magnitude of the second velocity can then be obtained, since its direction is known and the direction and magnitude of one of its rectangular components are known. A line is drawn through the terminus of this known component and perpendicular to it.

It is important to notice that the velocity vector in each case is the hypotenuse of a right triangle and that the two components form the right angle.

EXAMPLE 6-10.   COMPONENT METHOD

In Fig. 6-20, the velocity of point A is known, and the velocities of points B and C are required.

TO FIND V_B (Fig. 6-20a):

1. The velocity vector V_A is resolved into two rectangular components: one parallel to the line joining A and B and one perpendicular to this line.
2. The parallel component of V_B is then laid out equal to the parallel component of V_A.
3. The direction of the perpendicular component of V_B can then be drawn perpendicular to the parallel component and thus determines V_B.

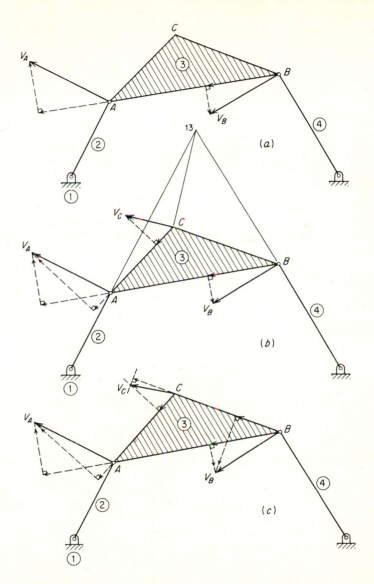

FIG. 6-20. Component method: four-link mechanism.

TO FIND V_C USING CENTRO 13 (Fig. 6-20*b*):

1. The velocity vector V_A is again resolved into two rectangular components, this time one parallel to the line joining 4 and C and one perpendicular to this line.
2. The parallel component of V_C is then laid out equal to the parallel component of V_A.
3. The direction of the perpendicular component of V_C can then be drawn perpendicular to the parallel component, and thus V_C can be determined. Notice that centro 13 provided the direction for V_C.

TO FIND V_C WITHOUT USING CENTRO 13 (Fig. 6-20*c*):

The only reason centro 13 is needed in Fig. 6-20*b* is to establish the direction of V_C. It is possible to obtain V_C without knowing its direction, by transferring components from both A and B. The procedure is as follows:

1. The velocity of B is obtained as before.
2. The velocity of A is resolved into two components, one parallel and one perpendicular to line AC, and the parallel component is transferred to C.
3. The velocity of B is resolved into two components, one parallel and one perpendicular to line BC, and the parallel component is transferred to C.
4. Through the terminus of each of the two components at C a line is drawn perpendicular to the component. The intersection of these two lines determines the velocity of point C.

EXAMPLE 6-11.   COMPONENT METHOD: SLIDER CRANK

In Fig. 6-21, V_A is given, and it is required to find V_B. The velocity of A is resolved into two rectangular components, one parallel and one perpendicular to link 3. The parallel component of V_B is then drawn equal

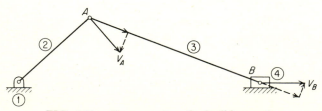

FIG. 6-21.   Component method: slider crank.

to the parallel component of V_A. The perpendicular component of V_B can then be drawn perpendicular to the parallel component and thus V_B can be determined.

6-7. Component Method Applied to Sliding Links

The general case of two sliding links where both links are moving requires somewhat different treatment. The solution of this type of problem is based on the fact that the normal velocity components of the two contact points must be equal or else the two links would be crushing or separating. In Fig. 6-22, the contact point is represented by E on link 2 and F on link 3. The velocity of point E is known, and it is required to find the velocity of point F. If both V_E and V_F were resolved into two components, one normal and one tangent to the point of contact, it is obvious that both of the normal components would have to be equal.

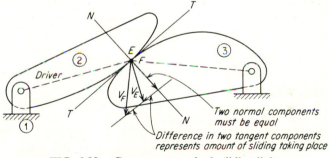

FIG. 6-22. Component method: sliding links.

Therefore, the first step is to resolve the known velocity V_E into two components, one normal to the contact point and one tangent. Then, since the direction of the unknown velocity is known (perpendicular to its radius) and its normal component is known (must be equal to the normal component of V_E), its magnitude can be found as shown.

The same reasoning applies to the two mechanisms in Fig. 6-23a, where link 2 is the driver, and in Fig. 6-23b, where link 4 is the driver. In both cases point A is considered a point on link 2 that coincides with point B, which is considered on link 4.

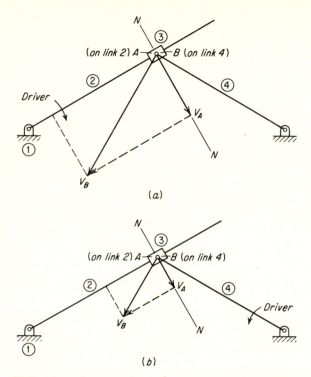

FIG. 6-23.   Component method: sliders on moving links.

Notice in Fig. 6-23a and b that one of the velocities is normal to the contacting surfaces and, therefore, has no tangential component.

6-8.   Proportionality of Perpendicular Components

It was shown in Fig. 6-19 that points along a rigid link must have the same component of velocity *parallel* to the link (that is, parallel to the line joining the points). It can also be shown that the *perpendicular* components of the velocities of points along a rigid link have useful proportionality characteristics; that is, if the perpendicular components of two points on a link, such as points A and B in Fig. 6-24, are known, a *line of proportion* can be drawn to determine the

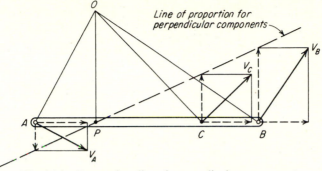

FIG. 6-24. Proportionality of perpendicular components.

perpendicular components of any other point along the link, such as *C*. This makes it very easy to find the velocity of any point along a floating link without locating the centro. It should be noted that point *P* in Fig. 6-24 is the point where the perpendicular component is zero. For this point to be correctly located, it is important that the perpendicular components be drawn with their origins at the respective points rather than at the termini of the parallel components as was done in Fig. 6-20.

PROOF

In Fig. 6-25, the velocities of *A*, *B*, and *C* have been resolved into components parallel and perpendicular to the link. The center of rotation of the link is at *O*, and *OA*, *OB*, and *OC* are the radii for points *A*, *B*, and *C*. Point *A* has no perpendicular component because its radius is perpendicular to the link.

Considering the velocity V_B, it is evident that the triangles *OAB* and *EFB* are similar because of mutually perpendicular sides, so

$$\frac{FB}{AB} = \frac{EF}{OA}$$

and since *EF/OA* is constant (along a rigid link),

$$FB \sim AB$$

In other words, *the perpendicular component of a velocity along a rigid link is proportional to its distance from the point that has no perpendicular component.*

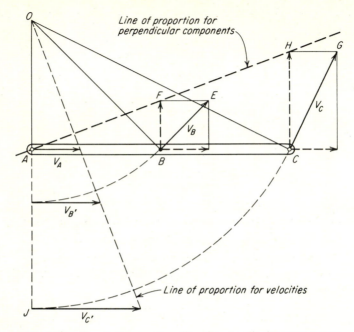

FIG. 6-25.   Proof of proportionality of perpendicular components.

In Art. 5-3 the concept of treating all instantaneous motion as pure rotation was explained. This concept is used throughout the text. It is possible, however, to treat instantaneous motion as a combination of rotation and translation. Figure 6-26a shows that the displacement of link 2 can be represented as pure rotation about point O. Figure 6-26b shows how the same displacement can be represented as a combination of translation s and rotation θ.

Referring again to Fig. 6-24, it should be noted that although the motion of the link was described as a rotation about point O, it could also be thought of as a rotation about point P and a translation along line AB. Notice then that all points along the link would have the same *translational* (parallel) component but that each point would have a different *rotational* (perpendicular) component proportional to its distance from the new center of rotation P.

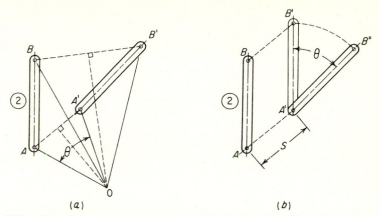

FIG. 6-26. Instantaneous motion of a link depicted as (*a*) pure rotation and (*b*) as a combination translation and rotation.

This concept of instantaneous motion is useful for applying the component method in situations where the velocities are required for several points along a floating link.

EXAMPLE 6-12. COMPONENT METHOD: USING LINE OF PROPORTION

In Fig. 6-27*a*, the velocity of point *A* is given, and the velocities of points *B*, *C*, and *D* are required.

SOLUTION

1. The velocity vector V_A is resolved into two rectangular components: one parallel to line *BACD* and one perpendicular to it.
2. The parallel component of V_C is then laid out equal to the parallel component of V_A. Since the direction of V_C is known (perpendicular to link 4), then V_C can be determined as shown.
3. The line of proportion can now be established for all perpendicular components along link 3 by first drawing the parallelograms at *A* and *C*, as shown in Fig. 6-27*b*. Now that the line of proportion is drawn locating point *P*, it is convenient to think of link 3 rotating about point *P* and translating along line *BACD*.
4. The velocities of points *B* and *D* are now readily obtained by adding their rotational components to their translational components by completing the parallelograms as shown.

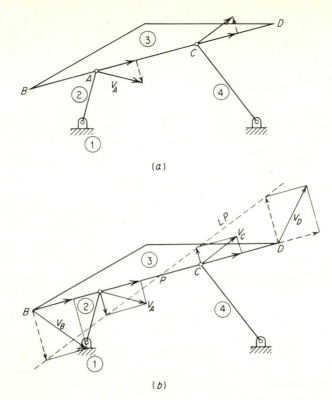

FIG. 6-27. Component method: using line of proportion.

Figure 6-28 shows how the combination rotation-and-translation concept applies to noncollinear points along a floating link. Notice that the link has its centro at O, but once the line of proportion is established, it is more convenient to think of the link's motion as a rotation about P and a translation along AB. The rotational component of D is obtained by rotating D to line AB (rotating about P) and using the line of proportion. Then rotate the component back to D. The velocity of D is obtained by adding the translational component (which is the same for all points on the link) to this rotational component.

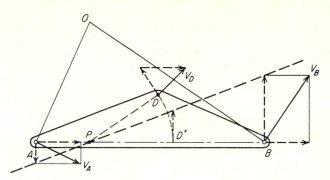

FIG. 6-28. Perpendicular component of a noncollinear point.

EXAMPLE 6-13. COMPONENT METHOD: USING LINE OF PROPORTION FOR NONCOLLINEAR POINT

In Fig. 6-29, the velocity of A is given, and the velocities of points B, C, and D are required.

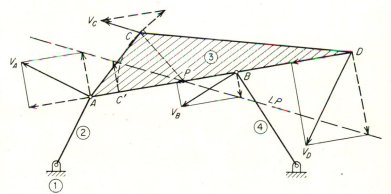

FIG. 6-29. Component method: using line of proportion for noncollinear point.

SOLUTION

1. The velocity of A is resolved into two components, one parallel and one perpendicular to line AB.

2. The parallel component of V_A is transferred to B. Since the direction of V_B is known and since the direction of its perpendicular component is known, V_B can be determined as shown.
3. The line of proportion for the perpendicular components is now established as shown. Once the line of proportion is drawn, it is convenient to think of link 3 rotating about point P and translating along line ABD.
4. The velocity of D is obtained by adding its rotational component (determined by the line of proportion) to its translational component (transferred from A). This addition is accomplished by drawing the parallelogram as shown.
5. Point C is rotated about P to line AB, where the magnitude of its rotational component is determined by the line of proportion, after which the component is rotated back to C. The translational component is added to this rotational component to obtain V_C.

6-9. The Relative-velocity Method

The velocity methods used so far are sufficient for solving problems involving *absolute* velocities of various points in a mechanism. When accelerations of mechanisms are studied (Chap. 7), however, it is necessary to determine the relative velocities between points on a link. For this reason, the relative-velocity method of obtaining velocities is the most useful method. The other methods, even though not so universally used, are invaluable as means of gaining insight into velocities and are convenient to fall back on occasionally in assisting a relative-velocity solution.

In addition to providing relative velocities, the relative-velocity method has other advantages. The construction work can be drawn to one side of the mechanism, thereby leaving the kinematic drawing uncluttered; the centros of floating links are not required; and the actual construction or drafting is an absolute minimum.

EXPLANATION OF METHOD

Considering the two moving points A and B in Fig. 6-30, the absolute velocity V_B of point B is equal to the vector sum of the

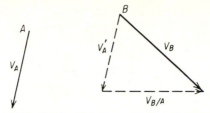

FIG. 6-30. Relative velocity of two points—general.

absolute velocity V_A of point A and the velocity $V_{B/A}$ of B relative to A. This is expressed in the vector equation

$$V_B = V_A \leftrightarrow V_{B/A} \qquad\qquad (3\text{-}19)$$

as was shown in Art. 3-15. In the *general* case, then, it is necessary to know both the velocity V_A of A and the velocity $V_{B/A}$ of B relative to A to find the velocity V_B of B. If the points A and B are on the same rigid link, however, the situation becomes simpler, for the *direction* of the relative-velocity vector is known, making it possible to find the velocity of B without knowing the magnitude of the relative velocity vector. The following statement, then, is the basis for the relative-velocity method:

STATEMENT

The only relative velocity (or motion) that can exist between two points on a rigid link is in a direction perpendicular to a line connecting the two points.

This statement should be self-evident. If there were *any* relative velocity (or motion) existing between two points along their connecting line, the link would be crushing, bending, stretching, or breaking. The concept is much easier to understand in the case of a pivoted link than it is in the case of a floating link. In Fig. 6-31, it is obvious that the only velocity (or motion) that B can have relative to A is, as shown, perpendicular to a line connecting the two points. In this case, the motion of B relative to A is the absolute motion of B since A is on the frame.

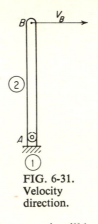

FIG. 6-31.
Velocity
direction.

On the basis of this statement it will be shown that, if the velocity of one point on a link is known and the *direction* of a second point is known, the velocity of the second point can be found. In Fig. 6-32a, the velocity of A is known and the direction only of B is known. A *velocity diagram* is constructed to one side as follows:

1. Arbitrarily locate the origin o (note lower-case letter), which represents a fixed point. All vectors originating at o represent absolute velocities.
2. The known absolute velocity V_A is laid off from o.
3. The *direction* of the absolute velocity V_B is drawn through o.
4. The direction of the relative velocity $V_{B/A}$ is known to be perpendicular to the line connecting A and B, and it must connect the termini of the two absolute velocities V_A and V_B (see Fig. 6-30). Therefore, a line is drawn through the known terminus of V_A and in a direction perpendicular to line AB.
5. This establishes the magnitudes of V_B and $V_{B/A}$.

Note that $V_{B/A}$ points toward V_B. A vector of opposite sense would represent $V_{A/B}$. Note also that the absolute velocities emanate from the origin o and that the terminus of each is labeled with a lower-case letter corresponding to the point involved. This conventional method of labeling the velocity diagram is most important, as will be seen in later examples.

From the velocity diagram in Fig. 6-32b it is evident that

$$V_B = V_A \mathbin{+\!\!\!\!\!\rightarrow} V_{B/A}$$

and
$$V_A = V_B \mathbin{+\!\!\!\!\!\rightarrow} V_{A/B}$$
(3-13)

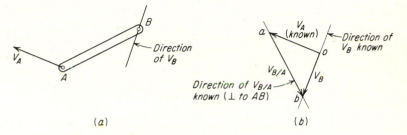

FIG. 6-32.   Relative-velocity method: two points on rigid link.

EXAMPLE 6-14.   RELATIVE-VELOCITY METHOD: DIRECTIONS OF POINTS ON FLOATING LINK KNOWN

In Fig. 6-33a, the velocity of A is known, and it is required to find the velocities of B, C, and D.

SOLUTION

The velocity diagram for this example is developed in steps for illustrative purposes. The steps shown in Fig. 6-33b, c, and d should actually be combined in one diagram as in Fig. 6-33e.

TO FIND V_B:

1. The given velocity V_A is laid off from the origin, as shown in Fig. 6-33b.
2. The direction of V_B (perpendicular to link 4) is drawn to indefinite length through the origin.
3. The direction of $V_{B/A}$ (perpendicular to line AB) is drawn through the terminus of V_A.
4. The intersection of the V_B and $V_{B/A}$ direction lines determines the magnitudes of both.

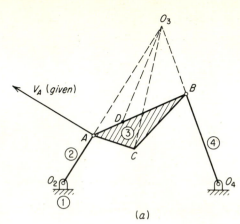

(a)

To find V_B :

V_A (given)

Direction of V_B known

$V_{B/A}$

V_B

Direction of $V_{A/B}$ known ($\perp$ to line AB)

(b)

To find V_C :

$V_{C/A}$

V_A (given)

V_C

Direction of V_C known

Direction of $V_{A/C}$ known ($\perp$ to line AC)

(c)

To find V_D :

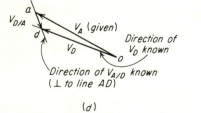

$V_{D/A}$

V_A (given)

V_D

Direction of V_D known

Direction of $V_{A/D}$ known ($\perp$ to line AD)

(d)

Complete diagram :

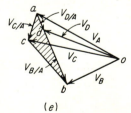

$V_{C/A}$ $V_{D/A}$ V_D V_A

V_C

$V_{B/A}$ V_B

(e)

FIG. 6-33. Relative-velocity method: using centro for floating link.

TO FIND V_C:

1. The given velocity is again laid off from the origin, as shown in Fig. 6-33c.
2. The direction of V_C (perpendicular to O_3C) is drawn to indefinite length through the origin.
3. The direction of $V_{C/A}$ (perpendicular to line AC) is drawn through the terminus of V_A.
4. The intersection of the V_C and $V_{C/A}$ direction lines determines the magnitudes of both.

TO FIND V_D:

1. The given velocity V_A is again laid off from the origin, as shown in Fig. 6-33d.
2. The direction of V_D (perpendicular to O_3D) is drawn to indefinite length through the origin.
3. The direction of $V_{D/A}$ (perpendicular to line AD) is drawn through the terminus of V_A.
4. The intersection of the V_D and $V_{D/A}$ direction lines determines the magnitudes of both.

VELOCITY-IMAGE CONCEPT

The usefulness of the velocity diagram is increased greatly if it is realized that the diagram forms a *velocity image* of the mechanism. In Fig. 6-33e, it is not difficult to recognize the shaded portion *abc* as being an image of link 3 (*ABC*) in the mechanism. The fact that the image was produced by drawing lines perpendicular to the corresponding sides of link 3 proves that the figures are similar (triangles with mutually perpendicular sides are similar).

In the same velocity diagram vector *oa* is the velocity image of link 2, vector *ob* is the velocity image of link 4, and the origin *o* can be considered the velocity image of link 1, the frame. This concept will be utilized in later examples.

EXAMPLE 6-15. RELATIVE-VELOCITY METHOD: DIRECTIONS OF POINTS ON FLOATING LINK NOT KNOWN

This example (Fig. 6-34) is exactly like the previous example except that the centro of the floating link 3 is not used to establish the directions of

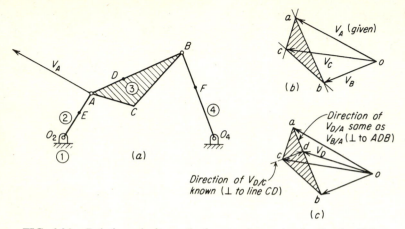

FIG. 6-34.   Relative-velocity method: not using centro for floating link.

C and *D* and two points *E* and *F* have been added. The velocity of *A* is known, and it is required to find the velocities of *B*, *C*, *D*, *E*, and *F*.

TO FIND V_B:

V_B is found exactly as in the previous example.

TO FIND V_C (Fig. 6-34*b*):

1. Draw a line through the terminus of V_A in the direction of $V_{C/A}$ ($\perp$ to *CA*).
$$V_C = V_A \mathbin{+\!\!\!\!\!\rightarrow} V_{C/A}$$

2. Draw a line through the terminus of V_B in the direction of $V_{C/B}$ ($\perp$ to *CB*).
$$V_C = V_B \mathbin{+\!\!\!\!\!\rightarrow} V_{C/B}$$

3. The terminus of V_C is at the intersection of these two lines, as shown in the upper velocity diagram (Fig. 6-34*b*). Determining the velocity of *C* in this manner is in effect solving the two simultaneous vector equations.

TO FIND V_D (Fig. 6-34*c*):

Since *A*, *D*, and *B* lie along a straight line on link 3, the relative velocities $V_{D/A}$ and $V_{D/B}$ cannot be used together to get V_D, because they are parallel (coincide) and will yield no intersection. Therefore, $V_{D/C}$ must be used in combination with $V_{D/A}$ (or $V_{D/B}$). This is done as follows:

1: Draw a line through the terminus of V_C in the direction of $V_{D/C}$ ($\perp$ to line DC).

$$V_D = V_C \overset{+}{\to} V_{D/C}$$

2. Draw a line through the terminus of V_A in the direction of $V_{D/A}$ ($\perp$ to line DA). This coincides with the $V_{B/A}$ line already drawn, since A, D, and B lie on a straight line.

$$V_D = V_A \overset{+}{\to} V_{D/A}$$

3. The terminus of V_D is established by the intersection of these two lines, as shown in the lower velocity diagram (Fig. 6-34c).

TO FIND V_D BY PROPORTION:

An easier way to locate V_D is to take advantage of the fact that the velocity diagram is an *image* of the actual mechanism, as shown in Fig.

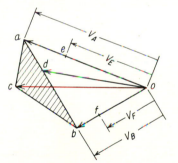

FIG. 6-35. Use of proportions in the velocity diagram.

6-35. Distances along the images are proportional to corresponding distances along the links. Therefore,

$$\frac{AD}{AB} = \frac{ad}{ab}$$

Similarly, the velocities of points E on link 2 and F on link 4 can be determined:

$$\frac{O_2E}{O_2A} = \frac{oe}{oa} \quad \text{and} \quad \frac{O_4F}{O_4B} = \frac{of}{ob}$$

EXAMPLE 6-16.   RELATIVE-VELOCITY METHOD: SLIDER CRANK

In Fig. 6-36, the velocity of A is known, and it is required to find the velocities of B and C.

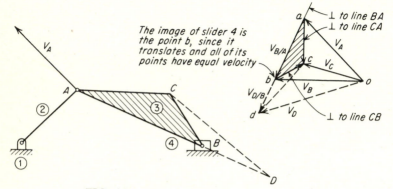

FIG. 6-36.   Relative-velocity method: slider crank.

TO FIND V_B (direction known):

1. Draw vector V_A in velocity diagram.
2. Lay off direction of V_B.
3. From terminus of V_A, lay off direction of $V_{B/A}$ ($\perp$ to line BA).
4. The intersection of $V_{B/A}$ and V_B lines determines the magnitudes of both ($V_B = V_A \mathbin{+\!\!\!\!\!>} V_{B/A}$).

TO FIND V_C:

1. Draw a line through the terminus of V_A in the direction of $V_{C/A}$.
2. Draw a line through the terminus of V_B in the direction of $V_{C/B}$.
3. The intersection of $V_{C/A}$ and $V_{B/A}$ determines V_C.

Steps 1 and 2 are based on the vector equations

$$V_C = V_A \mathbin{+\!\!\!\!\!>} V_{C/A} \qquad \text{and} \qquad V_C = V_B \mathbin{+\!\!\!\!\!>} V_{C/B}$$

In the above example, if it were required to find the velocity of a point such as D on link 3, it would be simple to extend the image in the velocity diagram by proportion, as shown in the figure ($AB/AD = ab/ad$).

EXAMPLE 6-17. RELATIVE-VELOCITY METHOD: ROLLING LINK

In Fig. 6-37, the velocity of A is known, and it is required to find the velocity of C.

It is assumed that no slipping occurs. Therefore, $V_A = V_B$, and the mechanism could be represented by the equivalent four-bar mechanism shown in dashed lines. Again, as in Example 6-5, it must be emphasized that the equivalent linkage shown in Fig. 6-37 is equivalent only for velocity analysis. Equivalent linkages for acceleration analysis are discussed in Chap. 7.

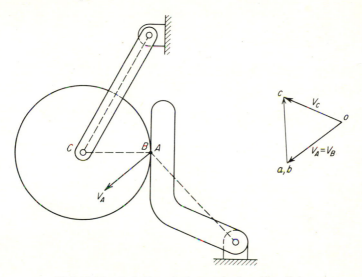

FIG. 6-37. Relative-velocity method: rolling link.

6-10. Relative-velocity Method Applied to Sliding Links

The solutions of sliding-link problems are based on the fact that the only possible relative velocity (or motion) that can exist between two sliding surfaces is along their common tangent line. This establishes the direction of relative motion, which provides sufficient information to complete the velocity diagram. Figure 6-38 shows two sliding-link problems solved.

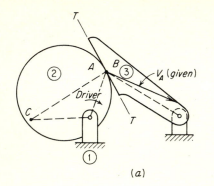

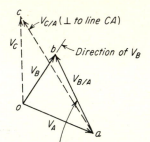

Only possible relative motion
between links 2 and 3 is along
common tangent T–T

(a)

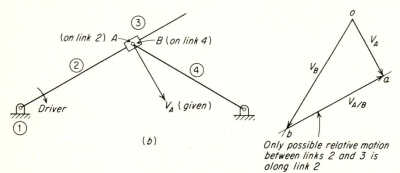

Only possible relative motion
between links 2 and 3 is
along link 2

(b)

FIG. 6-38.   Relative-velocity method: sliding links.

6-11.   Relative-velocity Method Applied to Complex Mechanisms

The application of the relative-velocity method to more complex mechanisms presents no particular difficulty if the directions of the various points are known.

EXAMPLE 6-18.   RELATIVE-VELOCITY METHOD: COMPLEX MECHANISM WHERE ALL DIRECTIONS ARE KNOWN

In Fig. 6-39a, the velocity of point A is given and the velocities of points B and C are required.

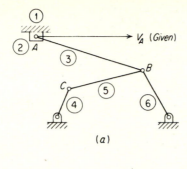

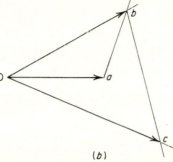

FIG. 6-39. Relative-velocity method in complex mechanism where all directions are known.

SOLUTION

1. Draw *oa* in the velocity diagram (Fig. 6-39*b*).
2. Lay out the direction of *ob* ($\perp$ to link 6).
3. Through *a*, lay out the direction of *ab* ($\perp$ to link 3).
4. The intersection of the *ob* and *ab* direction lines locates *b*.
5. Lay out the direction of *oc* ($\perp$ to link 4).
6. Through *b*, lay out the direction of *bc* ($\perp$ to link 5).
7. The intersection of the *oc* and *bc* direction lines locates *c*.

In analyzing complex mechanisms, the situation is sometimes encountered where the direction of one of the points on a floating link is not known. This results in a vector equation with too many unknowns to solve. In these cases, a trial-and-error approach can be used to produce a solution.

EXAMPLE 6-19.   RELATIVE-VELOCITY METHOD: COMPLEX MECHANISM
USING TRIAL-AND-ERROR APPROACH

In Fig. 6-40a, the velocity of point A is given and the velocities of points
B, C, and D are required. Notice that to find V_C the equation would be
$V_C = V_A \mathbin{+\!\!\!\!\rightarrow} V_{C/A}$. Neither the direction nor the magnitude of V_C is
known, nor is the magnitude of $V_{C/A}$ known. Since this is too many
unknowns to solve conventionally, the approach must be an indirect one.

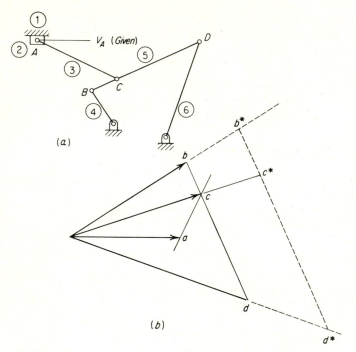

FIG. 6-40.   Relative-velocity method in complex mechanism
using trial-and-error approach.

SOLUTION

1. Draw *oa* in the velocity diagram (Fig. 6-40b).
2. Lay out the directions for *ob* and *oc*.
3. Choose a trial position for *b*, labeling it *b**.

4. Through b^*, lay out the direction for bd ($\perp$ to link 5) to locate d^*.
5. Lay out c^* in the same relative position along b^*d^* that C has relative to B and D, that is,

$$\frac{BC}{BD} = \frac{b^*c^*}{b^*d^*}$$

6. Now a line may be drawn from o to c^*, thereby establishing the direction of V_C.
7. Through a, draw the direction of line ac ($\perp$ to link 3). The intersection of this line and the line drawn in step 6 locates c.
8. Through c, draw a line parallel to b^*d^*, which will locate the correct positions for b and d.

Note: An alternative approach to this problem would be to determine the direction V_C by locating centro 15.

6-12. Summary of Velocity Methods

There are four basic graphical methods for obtaining velocities in mechanisms:

The *centro method*, which is quick and easy for simple mechanisms where all the centros needed are easily located and located within the immediate proximity of the drawing. This method is almost always used for obtaining velocities of various points on links rotating about fixed centers. This is a *direct* method, for the velocity on a particular link can be found directly without progressing through the intermediate links.

The *parallel-line method* provides a means of obtaining velocities on floating links even though their centro is inaccessible. This method is usually used in connection with the centro method. The disadvantage of this method is that velocities must be *chased through* the mechanism link by link to find a particular velocity.

The *component method* also provides a means of obtaining velocities on floating links even though their centro is inaccessible. This method is particularly suited to sliding-contact links but also has the disadvantage that velocities must be chased from link to link.

The *relative-velocity method* is the most universally used method and provides the most flexibility. Its advantages are:

1. It provides relative velocities as well as absolute velocities.
2. The velocity diagram does not clutter the actual drawing.
3. Velocities of points on floating links can be obtained without finding their centros.

The example problems that were shown in this chapter were somewhat unrealistic in that they all provided a known linear velocity on the crank and required the finding of some linear velocities on the other links. In most practical problems, the known velocity is in the form of the angular velocity of the driver. This must first be converted to a linear velocity of some point on the driver and then represented on the drawing by a vector drawn to some arbitrary scale. Quite often it is required to find the angular velocities of the other links in the mechanism rather than linear velocities of points. The example problem that follows is more typical of a practical problem.

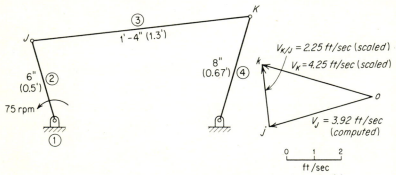

FIG. 6-41.   Relative velocities used for finding angular velocities.

EXAMPLE 6-20.   RELATIVE VELOCITIES USED FOR FINDING ANGULAR VELOCITIES

In Fig. 6-41, the angular velocity ω_2 of link 2 is known, and it is required to find the angular velocities of links 3 and 4.

$$\omega_2 = \frac{2\pi n}{60} = \frac{6.28 \times 75}{60} = 7.85 \text{ rad/sec}$$

$$V_J = r\omega = 0.5 \times 7.85 = 3.92 \text{ ft/sec}$$

V_J is then laid out to scale in the velocity diagram. Directions of V_K and $V_{K/J}$ are laid out in the velocity diagram, and their intersection determines the magnitudes of both. Then

$$\omega_3 = \frac{V_{K/J}}{r} = \frac{2.25}{1.3} = 1.73 \text{ rad/sec}$$

(see Art. 3-15, Example 3-1), or

$$n_3 = \frac{60\omega}{2\pi} = \frac{60 \times 1.73}{6.28} = 16.5 \text{ rpm}$$

$$\omega_4 = \frac{V_K}{r} = \frac{4.25}{0.67} = 6.35 \text{ rad/sec}$$

or

$$n_4 = \frac{60\omega}{2\pi} = \frac{60 \times 6.35}{6.28} = 60.7 \text{ rpm}$$

(see Art. 3-12).

Problems

CENTRO METHOD

6-1.   The velocity of point A in Fig. 6-42 is 15 ft/sec to the right. Find the velocities of points B and C. *Scale:* 1 in. = 10 ft/sec.

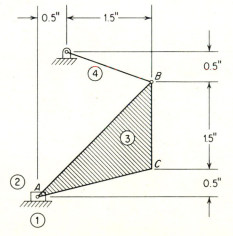

FIG. 6-42.   Probs. 6-1, 6-14, 6-24, and 6-36.

6-2. The velocity of point *A* in Fig. 6-43 is 20 ft/sec cw. (*a*) Use the link-to-link approach to find the velocities of points *B*, *C*, and *D*, in turn. (*b*) Obtain the velocity of point *D* by going directly from *A* to *B* (i.e., using the line of centers associated with links 1, 2, and 4). (*c*) Find the angular velocity of link 4 (including its sense). *Scale:* 1 in. = 20 ft/sec.

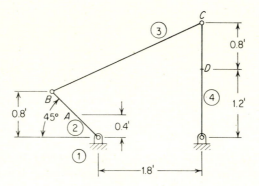

FIG. 6-43. Prob. 6-2.

6-3. The velocity of point *A* in Fig. 6-44 is 20 ft/sec cw. (*a*) Find the velocities of points *B*, *C*, and *D*. (*b*) Find the angular velocity of link 4. *Scale:* 1 in. = 20 ft/sec.

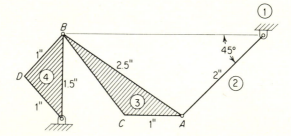

FIG. 6-44. Probs. 6-3, 6-15, 6-25, and 6-37.

6-4. The angular velocity of link 2 in Fig. 6-45 is 274 rad/sec cw. Find the velocities of points *A*, *B*, and *C*. *Scale:* 1 in. = 30 ft/sec.

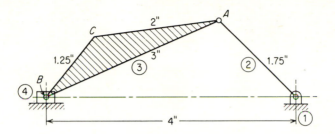

FIG. 6-45. Probs. 6-4, 6-16, 6-26, and 6-38.

6-5. The velocity of point *A* in Fig. 6-46 is 20 ft/sec cw. (*a*) Find the velocities of points *A*, *B*, *C*, *D*, and *E*. (*b*) Find the angular velocity of link 4. *Scale:* 1 in. = 20 ft/sec.

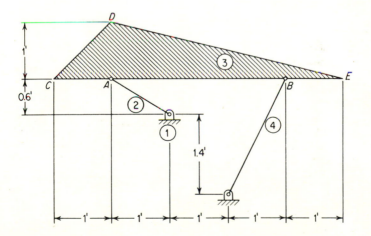

FIG. 6-46. Probs. 6-5, 6-17, 6-27, and 6-39.

6-6.   The velocity of point *A* in Fig. 6-47 is 10 ft/sec cw. (*a*) Find the velocity of point *B*. (*b*) Find the angular velocity of link 4. *Scale:* 1 in. = 10 ft/sec.

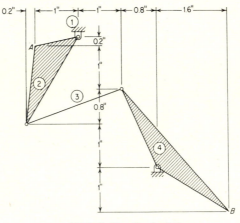

FIG. 6-47.   Prob. 6-6.

6-7.   The velocity of point *A* in Fig. 6-48 is 25 ft/sec ccw. (*a*) Find the velocities of points *B*, *C*, and *D*. (*b*) Find the angular velocity of link 4. *Scale:* 1 in. = 20 ft/sec.

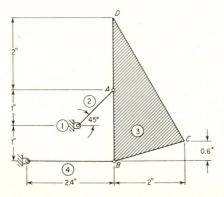

FIG. 6-48.   Probs. 6-7, 6-18, 6-28, and 6-40.

6-8.   The velocity of point *A* in Fig. 6-49 is 5 ft/sec ccw. Find the velocity of point *B*. *Scale:* 1 in. = 4 ft/sec.

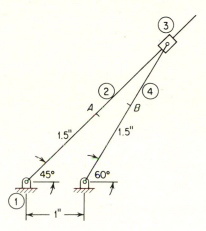

FIG. 6-49.   Probs. 6-8 and 6-29.

6-9.   The angular velocity of link 2 in Fig. 6-50 is 48 rad/sec cw. Find the angular velocity of link 3. *Scale:* 1 in. = 1 ft/sec.

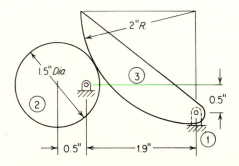

FIG. 6-50.   Probs. 6-9, 6-30, and 6-41.

6-10.   The velocity of point *A* in Fig. 6-51 is 30 ft/sec ccw. Find the velocities of points *B* and *C*. *Scale:* 1 in. = 20 ft/sec.

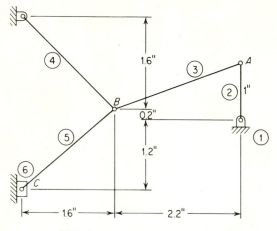

FIG. 6-51.   Probs. 6-10 and 6-47.

6-11.   The velocity of point *A* in Fig. 6-52 is 40 ft/sec ccw. Find the velocities of points *B*, *C*, and *D*. *Scale:* 1 in. = 40 ft/sec.

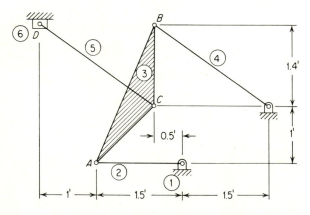

FIG. 6-52.   Probs. 6-11, 6-21, 6-33, and 6-48.

6-12.   The velocity of point *A* in Fig. 6-53 is 10 ft/sec ccw. Find the velocity of point *B*. *Scale:* 1 in. = 10 ft/sec.

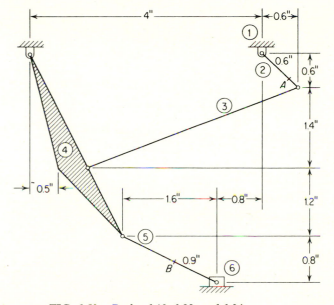

FIG. 6-53.   Probs. 6-12, 6-22, and 6-34.

6-13.   The angular velocity of link 2 in Fig. 6-54 is 257 rad/ sec cw. Find the velocities of points *A*, *B*, *C*, and *D*. *Scale:* 1 in. = 30 ft/sec.

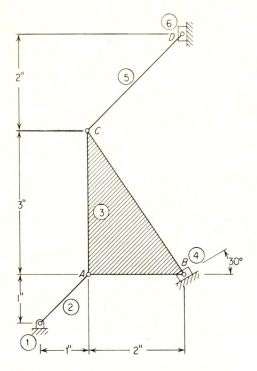

FIG. 6-54.   Probs. 6-13, 6-23, 6-35, and 6-49.

PARALLEL-LINE METHOD

6-14.　Same as Prob. 6-1 (Fig. 6-42).

6-15.　Same as Prob. 6-3 (Fig. 6-44).

6-16.　Same as Prob. 6-4 (Fig. 6-45).

6-17.　Same as Prob. 6-5 (Fig. 6-46).

6-18.　Same as Prob. 6-7 (Fig. 6-48).

6-19.　The velocity of point *A* in Fig. 6-55 is 10 ft/sec cw. Find the velocities of points *B*, *C*, *D*, and *E*. *Scale:* 1 in. = 10 ft/sec.

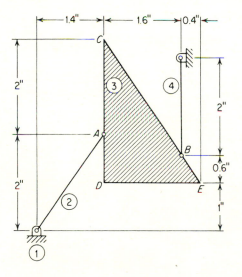

FIG. 6-55.　Prob. 6-19.

6-20.   The velocity of point *A* in Fig. 6-56 is 12 ft/sec cw. Find the velocities of points *B*, *C*, and *D*. *Scale:* 1 in. = 10 ft/sec.

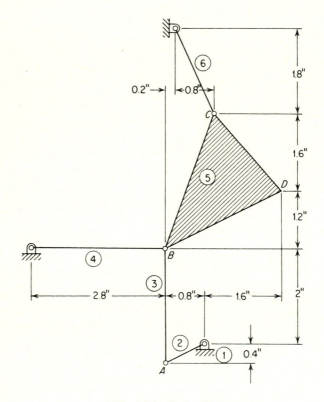

FIG. 6-56.   Prob. 6-20.

6-21.   Same as Prob. 6-11 (Fig. 6-52).
6-22.   Same as Prob. 6-12 (Fig. 6-53).
6-23.   Same as Prob. 6-13 (Fig. 6-54).

COMPONENT METHOD

6-24. Same as Prob. 6-1 (Fig. 6-42).
6-25. Same as Prob. 6-3 (Fig. 6-44).
6-26. Same as Prob. 6-4 (Fig. 6-45).
6-27. Same as Prob. 6-5 (Fig. 6-46).
6-28. Same as Prob. 6-7 (Fig. 6-48).
6-29. Same as Prob. 6-8 (Fig. 6-49).
6-30. Same as Prob. 6-9 (Fig. 6-50).

6-31.   The velocity of point *A* in Fig. 6-57 is 100 ft/sec cw. Find the velocities of points *B*, *C*, and *D*. *Scale:* 1 in. = 100 ft/sec.

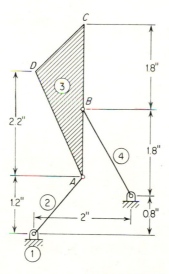

FIG. 6-57.   Prob. 6-31.

6-32.   The angular velocity of link 2 in Fig. 6-58 is 50 rad/sec cw. Find the velocities of points *A*, *B*, *C*, and *D*. *Scale:* 1 in. = 50 ft/sec.

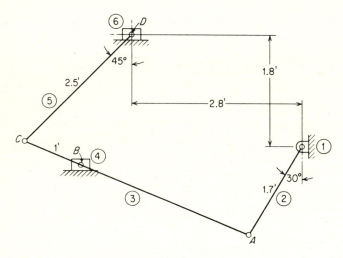

FIG. 6-58.   Prob. 6-32.

6-33.   Same as Prob. 6-11 (Fig. 6-52). *Scale:* 1 in. = 30 ft/sec.
6-34.   Same as Prob. 6-12 (Fig. 6-53).
6-35.   Same as Prob. 6-13 (Fig. 6-54).

RELATIVE-VELOCITY METHOD

Locational data as well as scales are given for the following problems to ensure that they will fit properly on an $8\frac{1}{2}$- by 11-in. sheet (with short edge horizontal). The locations for the link 2 pivot (or link 2 center in the case of a slider) and the origin of the velocity diagram are given in the form of full-scale coordinates from the left and bottom edges of the paper, respectively. Allowance is made for a $\frac{1}{4}$-in. margin on the top and right side, a $\frac{1}{2}$-in. margin on the left side, and a $\frac{3}{4}$-in. margin at the bottom.

6-36.   Same as Prob. 6-1 (Fig. 6-42) except also find the angular velocity of link 3. *Locational data:* slider 2 (2, 7); velocity diagram (2, 3). *Scale:* 1 in. = 5 ft/sec.

6-37.   Same as Prob. 6-3 (Fig. 6-44) except also find the angular velocity of link 3. *Locational data:* crank 2 (7, 10); velocity diagram (7, 4). *Scale:* 1 in. = 5 ft/sec.

6-38.   Same as Prob. 6-4 (Fig. 6-45) except also find the angular velocity of link 3. *Locational data:* crank 2 (7, 8); velocity diagram (2, 3). *Scale:* 1 in. = 10 ft/sec.

6-39.   Same as Prob. 6-5 (Fig. 6-46) except also find the angular velocity of link 3. *Locational data:* crank 2 (4, 8); velocity diagram (2, 4). *Scale:* 1 in. = 10 ft/sec.

6-40.   Same as Prob. 6-7 (Fig. 6-48) except also find the angular velocity of link 3. *Locational data:* crank 2 (2, $5\frac{1}{2}$); velocity diagram (7, $1\frac{1}{2}$). *Scale:* 1 in. = 10 ft/sec.

6-41.   Same as Prob. 6-9 (Fig. 6-50). *Locational data:* link 2 (4, 8); velocity diagram (4, 4). *Scale:* 1 in. = 1 ft/sec.

6-42.   The velocity of point *A* in Fig. 6-59 is 60 ft/sec cw. (*a*) Find the velocities of points *B* and *C*. (*b*) Find the angular velocities of links 3 and 4. *Locational data:* crank 2 (2, 8); velocity diagram (3, 6). *Scale:* 1 in. = 20 ft/sec.

6-43.   The velocity of point *A* in Fig. 6-60 is 40 ft/sec cw. (*a*) Find the velocities of points *B*, *C*, *D*, and *E*. (*b*) Find the angular velocities of links 3, 4, and 5. *Locational data:* crank 2 (6, $1\frac{1}{2}$); velocity diagram (4, 6). *Scale:* 1 in. = 20 ft/sec.

6-44.   The angular velocity of link 2 in Fig. 6-61 is 24 rad/ sec ccw. Find the velocities of points *A*, *B*, and *C*. *Locational data:* crank 2 (2, 3); velocity diagram (2, 7). *Scale:* 1 in. = 1 ft/sec.

6-45.   The angular velocity of link 2 in Fig. 6-62 is 48 rad/sec cw. (*a*) Find the velocities of points *A*, *B*, *C*, *D*, and *E*. (*b*) Find the angular velocities of links 3, 5, and 6. *Locational data:* crank 2 (2, 4); velocity diagram (7, $9\frac{1}{2}$). *Scale:* 1 in. = 3 ft/sec.

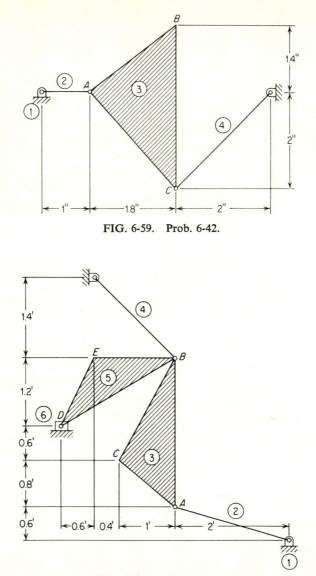

FIG. 6-59. Prob. 6-42.

FIG. 6-60. Prob. 6-43.

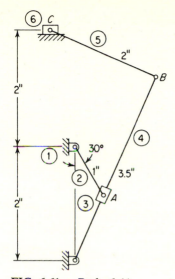

FIG. 6-61.   Prob. 6-44.

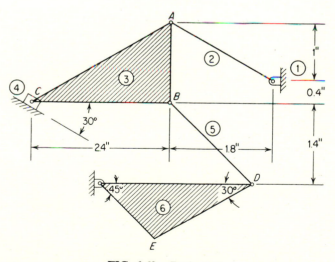

FIG. 6-62.   Prob. 6-54.

6-46. The velocity of point *A* in Fig. 6-63 is 30 ft/sec cw. (*a*) Find the velocities of points *B*, *C*, and *D*. (*b*) Find the angular velocities of links 3 and 5. *Locational data:* crank 2 (2, 10); velocity diagram (6, 6). *Scale:* 1 in. = 10 ft/sec.

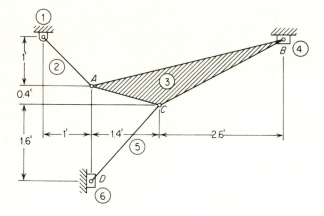

FIG. 6-63. Prob. 6-46.

6-47. Same as Prob. 6-10 (Fig. 6-51) except also find the angular velocity of link 5. *Locational data:* crank 2 (6, 9); velocity diagram (6, 6). *Scale:* 1 in. = 10 ft/sec.

6-48. Same as Prob. 6-11 (Fig. 6-52) except also find the angular velocities of links 3 and 4. *Locational data:* crank 2 (5, 7); velocity diagram (5, 6). *Scale:* 1 in. = 10 ft/sec.

6-49. Same as Prob. 6-13 (Fig. 6-54) except also find the angular velocities of links 3 and 5. *Locational data:* crank 2 (1, 4); velocity diagram (5, 5). *Scale:* 1 in. = 20 ft/sec.

6-50. The velocity of point *A* in Fig. 6-64 is 60 ft/sec cw. By the trial-and-error method (see Example 6-19) find the velocities of points

B, *C*, and *D*. *Locational data:* crank 2 (2, 9); velocity diagram (1, 6).
Scale: 1 in. = 20 ft/sec.

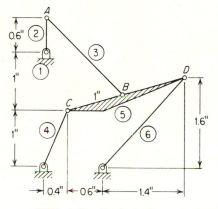

FIG. 6-64.   Prob. 6-50.

accelerations in mechanisms

7-1. Importance of Accelerations in Mechanisms

In actual mechanisms where links have mass, accelerations are important because the stresses in the links are often directly proportional to their accelerations ($F = ma$). In automotive or aircraft piston engines, the stresses imposed on the connecting rods and bearings that arise from accelerations are considerably greater than those which arise from the pressure of the burning gases.[1] A complete acceleration analysis of a mechanism is, then, fundamental to stress analysis and bearing design.

7-2. Relative-acceleration Method

The relative-acceleration method of analyzing accelerations of parts in a mechanism is based on the following principles:

1. That all motions are considered instantaneous
2. That the instantaneous motion of a point may be considered pure rotation
3. That the acceleration of a point is much more easily analyzed if it is resolved into two rectangular components, one normal and one tangent to its path

[1] J. Harland Billings, *Applied Kinematics*, 3d ed., p. 97, D. Van Nostrand Company, Inc., Princeton, N.J., 1953.

4. That the *relative* velocities as well as the *absolute* velocities of the various points in the mechanism are available. This requirement makes it most desirable to use the relative-velocity method to find the velocities involved.

The relative-acceleration method of determining accelerations in a mechanism is merely an extension of the relative-velocity method of obtaining velocities and utilizes an acceleration diagram that is very similar to the velocity diagram.

It was shown in Arts. 3-15 and 6-9 in discussing the velocities of two points *A* and *B* that the absolute velocity of point *B* is equal to the vector sum of the absolute velocity of point *A* and the velocity of *B* relative to *A*. This relative-velocity relationship was expressed by the vector equation $V_B = V_A + V_{B/A}$ [Eq. (3-19)]. Figure 7-1 repeats the illustration that was given in Fig. 6-30 showing the vector relationship.

It was also shown in Art. 3-15 that, if the velocity vectors shown in Fig. 7-1 were regarded as changes in velocity (ΔV) for a given infinitesimal interval of time (Δt), then the same vectors could represent accelerations ($a = \Delta V/\Delta t$) and the acceleration vectors would appear as shown in Fig. 7-2. This relative-acceleration relationship was expressed by the vector equation $a_B = a_A + a_{B/A}$ [Eq. (3-20)].

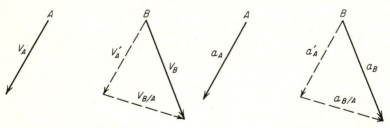

FIG. 7-1. Relative velocity.   FIG. 7-2. Relative acceleration.

ACCELERATION DIRECTIONS

It was stated in Art. 6-9 that the only relative velocity (or motion) that can exist between two points on a rigid link is in a direction perpendicular to a line connecting the two points. This principle provides the directions for all *relative* velocities, and, since the directions for most *absolute* velocities are obvious (perpendicular to their radii), the construction of the velocity diagram is comparatively simple.

Unfortunately, the construction methods for *acceleration* diagrams are not so straightforward. The directions of accelerations of points moving with curvilinear motion are not always known. It is necessary to regard accelerations as made up of two rectangular components: the normal component a^n and the tangential component a^t. This greatly simplifies the analysis, because the direction of these components is always known. As shown in Fig. 7-3, the normal

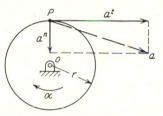

FIG. 7-3. Acceleration component directions—absolute acceleration.

component is directed toward the center of rotation, and the tangential component is perpendicular to the normal component, or tangent to the path of the point. It is evident that the acceleration is the vector sum of these two components; that is, that $a = a^n + a^t$ [Eq. (3-6)]. It should also be noted that, if the angular acceleration of the rotating body were zero, the tangential acceleration would be zero and a^n would equal a.

The situation is similar in the case of *relative* accelerations. Figure 7-4 shows a link with two points A and B, neither of which is fixed. If the link has an angular acceleration α, the normal and tangential acceleration components for the acceleration $a_{B/A}$ of point B relative to A are shown in Fig. 7-4a. The normal component $a^n_{B/A}$ points toward A, and the tangential component $a^t_{B/A}$ is perpendicular

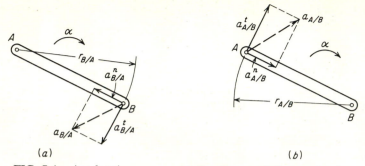

FIG. 7-4. Acceleration component directions—relative acceleration.

to the line AB. In Fig. 7-4b the components for the acceleration of point A relative to B are shown. In this case the normal component points toward B. Note that, in both these cases of relative accelerations, the distance between the two points involved is considered the radius. The magnitudes of these acceleration components may be expressed as follows:

$$a^n{}_{B/A} = \frac{(V_{B/A})^2}{r_{B/A}} \qquad a^n{}_{A/B} = \frac{(V_{A/B})^2}{r_{A/B}} \qquad (3\text{-}7)$$

and
$$a^t{}_{B/A} = r_{B/A}\alpha \qquad a^t{}_{A/B} = r_{A/B}\alpha \qquad (3\text{-}8)$$

RELATIVE-ACCELERATION EQUATION EXPANDED

The basic relative-acceleration equation $a_B = a_A \mathbin{+\!\!\!\!\!\gt} a_{B/A}$ illustrated in Fig. 7-2 is the basis for the acceleration diagram, but, since the directions of the various accelerations in the equation are not known, it is necessary to think of each term in the equation as being composed of its two rectangular components a^n and a^t, as shown in Figs. 7-3 and 7-4. Therefore, the basic relative-acceleration equation can be expanded as follows:

Since
$$a_B = a_A \mathbin{+\!\!\!\!\!\gt} a_{B/A} \qquad (3\text{-}20)$$

and
$$a = a^n \mathbin{+\!\!\!\!\!\gt} a^t \qquad (3\text{-}6)$$

then
$$a^n{}_B \mathbin{+\!\!\!\!\!\gt} a^t{}_B = a^n{}_A \mathbin{+\!\!\!\!\!\gt} a^t{}_A \mathbin{+\!\!\!\!\!\gt} a^n{}_{B/A} \mathbin{+\!\!\!\!\!\gt} a^t{}_{B/A} \qquad (7\text{-}1)$$

Equation (7-1) then becomes the basis for the construction of the acceleration diagram.

PATTERN OF SOLUTION

The solution of a typical acceleration problem involves two points on a rigid link, one whose acceleration is either known or immediately obtainable and the other whose acceleration is sought. Equation (7-1) is usually set up so that the left side of the equation represents the unknown acceleration. The fact that the two points involved are on the same rigid link simplifies the solution considerably, for the directions of all the vectors are known, as was illustrated in Figs. 7-3 and 7-4. Therefore, if the magnitudes of all but *two* of these vectors are known, the unknown acceleration can be determined.

It is important to realize that the two sides of the vector equation must add up to the same resultant—the unknown acceleration. If the directions of accelerations could be determined without breaking them into normal and tangential components, there would be only three vectors involved, and the acceleration diagram would appear as shown in Fig. 7-5a, resembling the velocity diagram. Since each acceleration must be resolved into two components, however, the

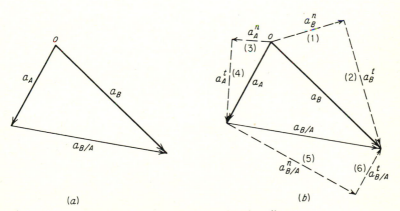

FIG. 7-5. Pattern of the acceleration diagram.

acceleration diagram usually resembles the vector arrangement shown in Fig. 7-5*b*. Notice that vectors 1 and 2 correspond to the left side of the basic equation [Eq. (7-1)] and that vectors 3, 4, 5, and 6 correspond to the right side. Notice also that the sum of either side is a_B, the acceleration being sought.

In a typical problem, the directions of all six components are known, the magnitudes of both the normal and tangential components of the *known* acceleration can be computed (a^n_A and a^t_A), and the magnitudes of the other two normal components can be computed (a^n_B and $a^n_{B/A}$). Magnitudes of the two remaining tangential components (a^t_B and $a^t_{B/A}$) cannot be computed for lack of data, but, since their directions are known, their intersection in the acceleration diagram establishes the terminus of the acceleration being sought (a_B).

In vector addition the order of the vectors does not generally matter, but in this case it is important that the pairs of normal and tangential components be adjacent so that the accelerations they represent can be readily obtained. Also, it is important that the last vector to be added for each side of the vector equation be the tangential component whose magnitude is unknown.

In many problems, certain terms of the basic equation are zero and will not appear in the vector diagram. For example, if the *known* point is on a link that rotates with constant angular velocity, it will have no tangential component, and its acceleration will equal its normal component.

EXAMPLE 7-1.   FOUR-LINK MECHANISM

In Fig. 7-6*a*, crank 2 is rotating counterclockwise (ccw) at 75 rpm and is slowing down at the rate of 15 rad/sec². It is required to find the accelerations of points *A* and *B* and the angular velocities and accelerations of links 3 and 4.

SOLUTION

1.   Draw the velocity diagram as shown in Fig. 7-6*b*.

$$\omega_2 = \frac{2\pi n}{60} = \frac{6.28 \times 75}{60} = 7.85 \text{ rad/sec}$$

$$V_A = r_A\omega_2 = 1.2 \times 7.85 = 9.4 \text{ ft/sec}$$

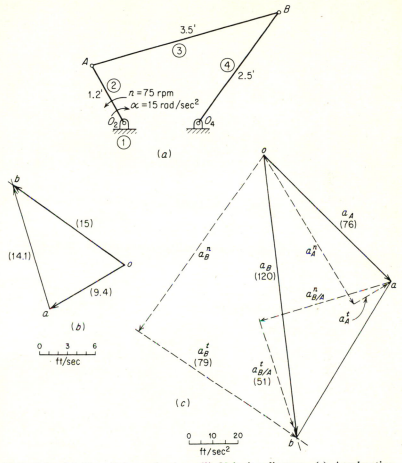

FIG. 7-6. (*a*) Four-link mechanism. (*b*) Velocity diagram. (*c*) Acceleration diagram.

a. Lay out V_A from the origin *o* ($\perp$ to link 2).

b. Lay out the *direction* of V_B from the origin *o* ($\perp$ to link 4).

c. Lay out the *direction* of $V_{B/A}$ from the terminus of V_A ($\perp$ to link 3).

d. The intersection of the V_B and $V_{B/A}$ direction lines determines the magnitudes of both.

2. Write the acceleration equation for a_B:

$$a_B = a_A \mathbin{+\!\!\!\!\rightarrow} a_{B/A}$$
$$a^n{}_B \mathbin{+\!\!\!\!\rightarrow} a^t{}_B = a^n{}_A \mathbin{+\!\!\!\!\rightarrow} a^t{}_A \mathbin{+\!\!\!\!\rightarrow} a^n{}_{B/A} \mathbin{+\!\!\!\!\rightarrow} a^t{}_{B/A}$$

3. Determine the magnitudes and directions of the various terms:

$$a^n{}_B = \frac{V_B{}^2}{r_B} = \frac{(15)^2}{2.5} = \frac{225}{2.5} = 90 \text{ ft/sec}^2 \qquad (\parallel \text{ to link 4})$$

$$a^t{}_B = r_B \alpha_4 \qquad \textit{direction known only } (\perp \text{ to link 4})$$

$$a^n{}_A = \frac{V_A{}^2}{r_A} = \frac{(9.4)^2}{1.2} = \frac{88.4}{1.2} = 73.7 \text{ ft/sec}^2 \qquad (\parallel \text{ to link 2})$$

$$a^t{}_A = r_A \alpha_2 = 1.2 \times 15 = 18 \text{ ft/sec}^2 \qquad (\perp \text{ to link 2})$$

$$a^n{}_{B/A} = \frac{(V_{B/A})^2}{r_{B/A}} = \frac{(14.1)^2}{3.5} = 56.8 \text{ ft/sec}^2 \qquad (\parallel \text{ to link 3})$$

$$a^t{}_{B/A} = r_{B/A} \alpha_3 \qquad \textit{direction known only } (\perp \text{ to link 3})$$

4. Perform the vector addition in the acceleration diagram to obtain a_B (Fig. 7-6c).
   a. Lay out $a^n{}_B$ from the origin o ($\parallel$ to link 4).
   b. Through the terminus of $a^n{}_B$ draw a perpendicular line of indefinite length representing the direction of $a^t{}_B$, whose magnitude is unknown.
   c. Again starting from the origin o, lay out $a^n{}_A$ ($\parallel$ to link 2).
   d. From the terminus of $a^n{}_A$ and perpendicular to it, lay out $a^t{}_A$. This establishes the acceleration a_A of A, which represents the *known* acceleration and scales 76 ft/sec².
   e. From the terminus of a_A, lay out $a^n{}_{B/A}$ ($\parallel$ to link 3).
   f. Through the terminus of $a^n{}_{B/A}$ and perpendicular to it, draw a line of indefinite length representing the direction of $a^t{}_{B/A}$, whose magnitude is unknown.
   g. The intersection of the two indefinite-length lines drawn in steps (b) and (f), representing $a^t{}_B$ and $a^t{}_{B/A}$, respectively, determines a_B, which scales 120 ft/sec². Notice that all the components corresponding to terms of the equation are drawn with dashed lines. The solid acceleration vectors may be added later.

5. Compute the angular velocities and accelerations:

$$\omega_3 = \frac{V_{B/A}}{r_{B/A}} = \frac{14.1}{3.5} = 4.03 \text{ rad/sec} \qquad \text{ccw} \qquad (V_{B/A} \text{ scaled})$$

$$\alpha_3 = \frac{a^t_{B/A}}{r_{B/A}} = \frac{51}{3.5} = 14.6 \text{ rad/sec}^2 \qquad \text{cw} \qquad (a^t_{B/A} \text{ scaled})$$

$$\omega_4 = \frac{V_B}{r_B} = \frac{15}{2.5} = 6 \text{ rad/sec} \qquad \text{ccw} \qquad (V_B \text{ scaled})$$

$$\alpha_4 = \frac{a^t_B}{r_B} = \frac{79}{2.5} = 31.6 \text{ rad/sec}^2 \qquad \text{cw} \qquad (a^t_B \text{ scaled})$$

7-3. Proportionality of Accelerations

It is important to realize that the acceleration of a point on a rotating body is proportional to its radius. The expressions for the normal and tangential components of the acceleration of a point on a rotating body ($a^n = r\omega^2$ and $a^t = r\alpha$) show that these acceleration components are also proportional to the radius of the point. This proportionality of accelerations and their components is shown graphically in Fig. 7-7.

It should be pointed out, however, that in the case of floating links, the instantaneous center for accelerations is not the same as the instantaneous center (centro) for velocities. Since the method of finding instantaneous centers for accelerations is not as straightforward as is the method for finding instantaneous centers for velocities, and since the acceleration-analysis methods in this book do not utilize this concept, it will not be discussed. Graphical methods for locating acceleration centers are presented in advanced texts in kinematics.[1-3]

EXAMPLE 7-2. SLIDER CRANK

In Fig. 7-8a, crank 2 rotates counterclockwise with a uniform angular velocity of 30 rad/sec. It is required to find the linear accelerations of points *A*, *B*, *C*, and *D* and the angular velocity and acceleration of link 3.

[1] J. E. Shigley, *Kinematic Analysis of Mechanisms*, pp. 147–149, McGraw-Hill Book Company, New York, 1969.

[2] N. Rosenauer and A. H. Willis, *Kinematics of Mechanisms*, pp. 145–156, Associated General Publications, Sydney, Australia, 1953; republished by Dover Publications, Inc., New York, 1967.

[3] K. Hain, *Applied Kinematics*, 2d ed., pp. 155–158, McGraw-Hill Book Company, New York, 1967.

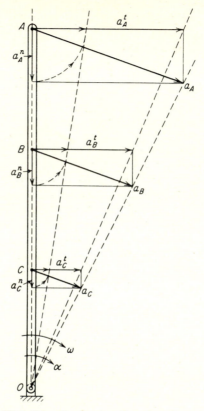

FIG. 7-7.   Proportionality of accelerations and their components.

SOLUTION

1.   Draw the velocity diagram as shown in Fig. 7-8b.

$$V_A = r_A \omega_2 = (\tfrac{3}{12})(30) = 7.5 \text{ ft/sec} \qquad (\perp \text{ to link 2})$$

$$V_B = \textit{direction known only} \qquad (\parallel \text{ to path of link 4})$$

$$V_{B/A} = \textit{direction known only} \qquad (\perp \text{ to } BA)$$

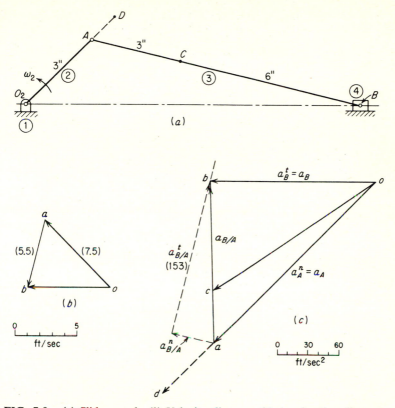

FIG. 7-8. (a) Slider crank. (b) Velocity diagram. (c) Acceleration diagram.

2. Write the acceleration equation for a_B:

$$a_B = a_A \overset{+}{\rightarrow} a_{B/A}$$

$$a_B = a^n_A \overset{+}{\rightarrow} a^t_A \overset{+}{\rightarrow} a^n_{B/A} \overset{+}{\rightarrow} a^t_{B/A}$$

Notice that a_B is not expanded into its components since B is on a slider. If it were resolved, the normal component would be equal to zero since $a^n = V^2/r = V^2/\infty = 0$, and the tangential component would be indeterminate since $a^t = r\alpha = \infty \cdot 0$. Therefore, since the normal component is equal to zero, $a_B = a^t_B$; and since a^t_B is indeterminate, only its direction is known—parallel to the path of the slider.

3. Determine the magnitudes and directions of the various terms.

$a_B = $ *direction known only*   ($\parallel$ to **path of slider**)

$$a^n_A = \frac{v_A^2}{r_A} = \frac{(7.5)^2}{^3\!/_{12}} = 225 \text{ ft/sec}^2 \qquad (\parallel \text{ to link 2})$$

$$a^t_A = r_A\alpha_2 = (^3\!/_{12})(0) = 0$$

$$\therefore a_A = a^n_A$$

$$a^n_{B/A} = \frac{(V_{B/A})^2}{r_{B/A}} = \frac{(5.5)^2}{^9\!/_{12}} = 40.3 \text{ ft/sec}^2 \qquad (\parallel \text{ to link 3})$$

$$a^t_{B/A} = r_{B/A}\alpha_3 \qquad \textit{direction known only } (\perp \text{ to link 3})$$

4. Perform the vector addition in the acceleration diagram (Fig. 7-8c) to obtain a_B.
   a. Lay out the direction of a_B ($a_B = a^t_B$) through the origin o.
   b. Again, from the origin, lay out a_A ($a_A = a^n_A$).
   c. From the terminus of a_A, lay out $a^n_{B/A}$.
   d. Through the terminus of $a^n_{B/A}$ and perpendicular to it, lay out the direction of $a^t_{B/A}$. The intersection of this line with the a_B direction line drawn in step a determines a_B.

5. Obtain a_C and a_D (Fig. 7-8c). Points c and d in the acceleration diagram are located by proportion, utilizing the *image concept* in a manner comparable with the velocity-image concept discussed in Art. 6-9. Point c is located on the image of link 3 so that $ac/ab = AC/AB$, and point d is located on the image of link 2 so that $od/oa = O_2D/O_2A$.

6. Determine the angular velocity and acceleration of link 3:

$$\omega_3 = \frac{V_{B/A}}{r_{B/A}} = \frac{5.5}{^9\!/_{12}} = 7.3 \text{ rad/sec} \qquad \text{cw}$$

$$\alpha_3 = \frac{a^t_{B/A}}{r_{B/A}} = \frac{153}{^9\!/_{12}} = 204 \text{ rad/sec}^2 \qquad \text{ccw}$$

EXAMPLE 7-3.   FOUR-LINK MECHANISM WITH EXTRA POINT ON FLOATING LINK

In Fig. 7-9a, crank 2 is rotating clockwise at 100 rpm and is speeding up at the rate of 120 rad/sec². It is required to find the linear accelerations

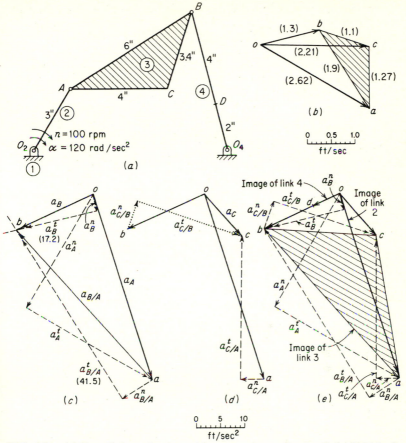

FIG. 7-9. (*a*) Four-link mechanism with extra point on floating link. (*b*) Velocity diagram. (*c*) to (*e*) Acceleration diagrams.

of points A, B, and C and the angular velocities and accelerations of links 3 and 4.

SOLUTION

For illustrative purposes, the acceleration diagram for this example is shown in several parts. In actual problem solving, it is usually more effective to show all the acceleration vectors on one diagram.

1. Draw the velocity diagram (Fig. 7-9b).

$$\omega_2 = \frac{2\pi n}{60} = \frac{6.28 \times 100}{60} = 10.5 \text{ rad/sec}$$

$$V_A = r_A\omega_2 = (\tfrac{3}{12})(10.5) = 2.62 \text{ ft/sec}$$

$V_B \qquad (\perp \text{ to link 4})$

$V_{B/A} \qquad (\perp \text{ to } BA)$

$V_{C/A} \qquad (\perp \text{ to } CA)$

$V_{C/B} \qquad (\perp \text{ to } CB)$

2. Write the acceleration equation for a_B:

$$a_B = a_A \overset{+}{\rightarrow} a_{B/A}$$
$$a^n{}_B \overset{+}{\rightarrow} a^t{}_B = a^n{}_A \overset{+}{\rightarrow} a^t{}_A \overset{+}{\rightarrow} a^n{}_{B/A} \overset{+}{\rightarrow} a^t{}_{B/A}$$

3. Determine the magnitudes and directions of the various terms.

$$a^n{}_B = \frac{V_B{}^2}{r_B} = \frac{(1.3)^2}{\tfrac{6}{12}} = 3.38 \text{ ft/sec}^2 \qquad (\| \text{ to link 4})$$

$$a^t{}_B = r_B\alpha_4 \qquad \textit{direction known only } (\perp \text{ to link 4})$$

$$a^n{}_A = \frac{V_A{}^2}{r_A} = \frac{(2.62)^2}{\tfrac{3}{12}} = 27.44 \text{ ft/sec}^2 \qquad (\| \text{ to link 2})$$

$$a^t{}_A = r_A\alpha_2 = (\tfrac{3}{12})(120) = 30 \text{ ft/sec}^2 \qquad (\perp \text{ to link 2})$$

$$a^n{}_{B/A} = \frac{(V_{B/A})^2}{r_{B/A}} = \frac{(1.9)^2}{\tfrac{6}{12}} = 7.22 \text{ ft/sec}^2 \qquad (\| \text{ to } BA)$$

$$a^t{}_{B/A} = r_{B/A}\alpha_3 \qquad \textit{direction known only } (\perp \text{ to } BA)$$

4. Perform the vector addition in the acceleration diagram to obtain a_B (Fig. 7-9c).

   a. Lay out $a^n{}_B$ from the origin o ($\|$ to link 4).
   b. Through the terminus of $a^n{}_B$ draw a perpendicular line of indefinite length representing the direction of $a^t{}_B$.
   c. Again starting from the origin o, lay out $a^n{}_A$ ($\|$ to link 2).
   d. From the terminus of $a^n{}_A$ and perpendicular to it, lay out $a^t{}_A$. This establishes a_A.
   e. From the terminus of a_A, lay out $a^n{}_{B/A}$ ($\|$ to BA).
   f. Through the terminus of $a^n{}_{B/A}$ and perpendicular to it, draw a line of indefinite length representing the direction of $a^t{}_{B/A}$. The intersection of this line with the $a^t{}_B$ direction line drawn in step b determines a_B.

5. Write the acceleration equation for a_C. There are two possibilities, since a_C can be written in terms of either a_A or a_B, both of which are now known. That is,

$$a_C = a_A +\!\!\!\!\!\to a_{C/A} \qquad \text{or} \qquad a_C = a_B +\!\!\!\!\!\to a_{C/B}$$

Expanding the first equation,

$$a_C = a_A +\!\!\!\!\!\to a^n{}_{C/A} +\!\!\!\!\!\to a^t{}_{C/A}$$

6. Determine the magnitudes and directions of the two unknown terms.

$$a^n{}_{C/A} = \frac{(V_{C/A})^2}{r_{C/A}} = \frac{(1.27)^2}{4/12} = 4.83 \text{ ft/sec}^2 \qquad (\parallel \text{ to } CA)$$

$$a^t{}_{C/A} = r_{C/A}\alpha_3 = r_{C/A}\left(\frac{a^t{}_{B/A}}{r_{B/A}}\right) = (4/12)\left(\frac{41.5}{6/12}\right)$$

$$= 27.8 \text{ ft/sec}^2 \qquad (\perp \text{ to } CA)$$

7. Perform the vector addition in the acceleration diagram to obtain a_C. Figure 7-9d shows how $a^n{}_{C/A}$ and $a^t{}_{C/A}$ are added to the known vector a_A to obtain a_C. It should be noted that *either* of the two equations for a_C could have been used. In fact, they may *both* be used, in which case it is not necessary to compute the magnitudes of the tangential components; their directions alone provide an intersection to establish the terminus of a_C, as shown with dotted lines in Fig. 7-9d. This in effect is solving two vector equations simultaneously.

In the complete acceleration diagram shown in Fig. 7-9e it is apparent that the terminus of a_C can also be located by completing the image of link 3 in the acceleration diagram by proportion ($AC/AB = ac/ab$ and $BC/AB = bc/ab$).

Similarly, the acceleration of point D on link 4 can be obtained by locating point d in the acceleration diagram by proportion ($O_4D/O_4B = od/ob$).

8. Determine the angular velocities and accelerations required:

$$\omega_3 = \frac{V_{B/A}}{r_{B/A}} = \frac{1.9}{6/12} = 3.8 \text{ rad/sec} \qquad \text{ccw}$$

$$\alpha_3 = \frac{a^t{}_{B/A}}{r_{B/A}} = \frac{41.5}{6/12} = 83 \text{ rad/sec}^2 \qquad \text{ccw}$$

$$\omega_4 = \frac{V_B}{r_B} = \frac{1.3}{6/12} = 2.6 \text{ rad/sec} \qquad \text{cw}$$

$$\alpha_4 = \frac{a^t{}_B}{r_B} = \frac{17.2}{6/12} = 34.4 \text{ rad/sec}^2 \qquad \text{ccw}$$

The acceleration vector equation (7-1) used in the preceding examples was developed for two points on the same rigid link. Occasionally, it is necessary to work with two points not on the same rigid link, which requires somewhat different treatment. One of three general methods may be used for handling such problems:

1. Draw an *equivalent linkage* whereby the given linkage is converted into a more conventional pin-jointed linkage.
2. Use *Coriolis's law*, which involves an additional acceleration component in the acceleration equation.
3. Construct a set of *motion curves* (displacement, velocity, and acceleration) for the particular point of interest in the mechanism.

The first two methods are discussed in the next two articles of this chapter; the third method is discussed in Chap. 8.

7-4. Equivalent Linkages

The acceleration analysis of many rolling (and sliding) contact mechanisms can be greatly simplified by constructing *equivalent linkages*, which convert particular instantaneous phases of such mechanisms to pin-jointed equivalents where the *known* and *unknown* points of interest are on the same rigid link.

Figure 7-10 shows several such mechanisms with their equivalent linkages shown with dashed lines. Notice in each case that the floating link of the equivalent linkage is drawn along the common normal of the two contacting surfaces and extends to the center of curvature of each of the surfaces. This ensures that points *A* and *B* at the ends of the equivalent link 3' (Fig. 7-10*a* to *i*) remain a constant distance apart for a finite period of time. This in turn ensures that the two pivoted links of the equivalent mechanism (2' and 4') will have the same angular-velocity ratio as the two pivoted links of the original mechanism (2 and 4) and can, therefore, be substituted for purposes of velocity and acceleration analyses.

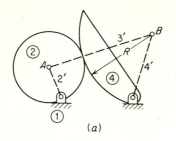
(a)

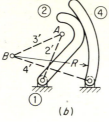
(b)

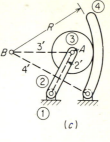
(c)

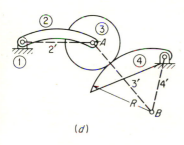
(d)

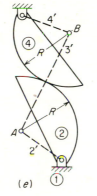
(e)

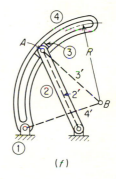
(f)

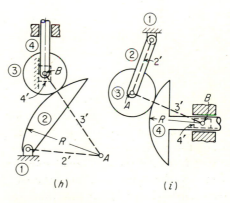
(g)

(h)

(i)

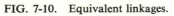

FIG. 7-10.   Equivalent linkages.

Figure 7-11 shows some mechanisms that *cannot* be handled by the equivalent-linkage method because one of the contacting links is straight or flat at the point of contact and its corresponding center of curvature is at infinity. Mechanisms of this type must be handled either by Coriolis's law (discussed in the next article) or by constructing motion curves (discussed in the next chapter).

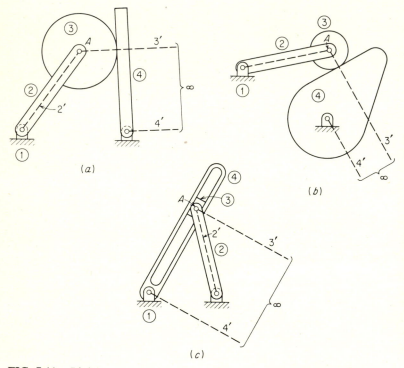

FIG. 7-11.   Linkages that cannot be worked by the equivalent-linkage method.

EXAMPLE 7-4.   CAM AND FOLLOWER: EQUIVALENT-LINK METHOD

In Fig. 7-12 the cam, link 2, is rotating counterclockwise with a constant angular velocity of 120 rad/sec. It is required to find the angular velocity and acceleration of link 3.

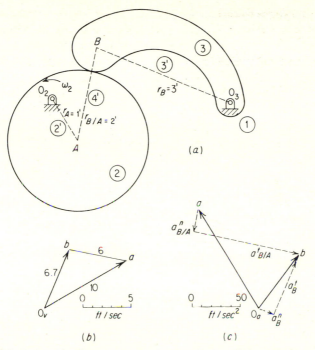

FIG. 7-12. Circular cam mechanism: equivalent-link method. (*a*) Circular cam. (*b*) Velocity diagram. (*c*) Acceleration diagram.

SOLUTION

1. Construct the *equivalent linkage* as shown by the dashed lines.
2. Draw the velocity diagram as shown in Fig. 7-12*b*.

$$V_A = r_A \omega_2 = (\tfrac{1}{12})(120) = 10 \text{ ft/sec} \qquad (\perp \text{ to link } 2')$$

$$V_B \qquad (\perp \text{ to link } 3')$$

$$V_{B/A} \qquad (\perp \text{ to link } 4')$$

3. Write the acceleration equation for a_B.

$$a_B = a_A \overset{+}{\rightarrow} a_{B/A}$$

$$a^n{}_B \overset{+}{\rightarrow} a^t{}_B = a^n{}_A \overset{+}{\rightarrow} a^t{}_A \overset{+}{\rightarrow} a^n{}_{B/A} \overset{+}{\rightarrow} a^t{}_{B/A}$$

4. Determine the magnitudes and directions of the various terms.

$$a^n_B = \frac{V_B^2}{r_B} = \frac{(6.7)^2}{3/12} = 179.6 \text{ ft/sec}^2 \qquad (\parallel \text{ to link } 3')$$

a^t_B *direction known only* ($\perp$ to link $3'$)

$$a^n_A = \frac{V_A^2}{r_A} = \frac{(10)^2}{1/12} = 1200 \text{ ft/sec}^2 \qquad (\parallel \text{ to link } 2')$$

$$a^t_A = 0 \qquad (\text{since } \alpha_2 = 0)$$

$$\therefore a_A = a^n_A$$

$$a^n_{B/A} = \frac{(V_{B/A})^2}{r_{B/A}} = \frac{(6)^2}{2/12} = 216 \text{ ft/sec}^2 \qquad (\parallel \text{ to link } 4')$$

$a^t_{B/A}$ *direction known only* ($\perp$ to link $4'$)

5. Perform the vector addition in the acceleration diagram as shown in Fig. 7-12c.
6. Calculate the angular velocity and angular acceleration for link $3'$ (which is the same as for link 3).

$$\omega_3 = \frac{V_B}{r_B} = \frac{6.7}{3/12} = 26.8 \text{ rad/sec} \qquad \text{cw}$$

$$\alpha_3 = \frac{a^t_B}{r_B} = \frac{70}{3/12} = 280 \text{ rad/sec}^2 \qquad \text{cw}$$

The preceding example can also be worked by using coincident points, which results in a Coriolis component of acceleration. See Example 7-6 (Fig. 7-18).

SCOTCH-YOKE MECHANISM

The scotch-yoke mechanisms shown in Fig. 7-13a and b are special cases of the mechanisms shown in Fig. 7-10h and i. They are equivalent to a slider-crank mechanism having an infinitely long connecting rod. Their equivalent linkage, shown in Fig. 7-13c, involves two coincident points A and B that are *not* on the same rigid link. Ordinarily, a problem involving coincident points that cannot be converted to an equivalent linkage involving two noncoincident points on the same link cannot be worked by the methods learned so

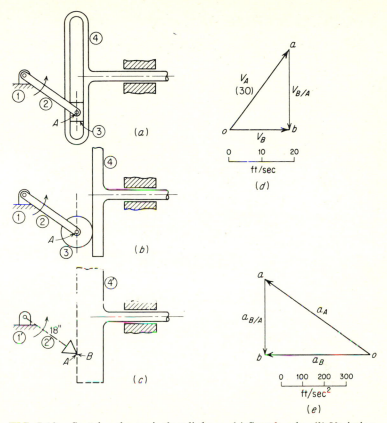

FIG. 7-13. Scotch-yoke equivalent linkage. (*a*) Scotch yoke. (*b*) Variation of scotch yoke. (*c*) Equivalent linkage. (*d*) Velocity diagram. (*e*) Acceleration diagram.

far. However, the fact that link 4 is not rotating makes the scotch yoke a special case that can be worked with the same basic equation (7-1) used in the preceding examples in this chapter. If link 4 rotated instead of translated, the problem would involve Coriolis acceleration and would have to be worked by the methods discussed in the next article.

EXAMPLE 7-5.   SCOTCH-YOKE MECHANISM

In Fig. 7-13a and b, the crank 2 is rotating counterclockwise with a constant angular velocity of 20 rad/sec. It is required to find the acceleration of the follower (link 4).

SOLUTION

1. Construct the *equivalent linkage* as shown in Fig. 7-13c. Since point A on link 2 traces a path along the dashed line shown, the pointed link 2′ moving along link 4′ extended is kinematically equivalent to the original linkage.

2. Draw the velocity diagram as shown in Fig. 7-13d.

$$V_A = r_A \omega_2 = (1.5)(20) = 30 \text{ ft/sec}$$
$$V_B \qquad (\parallel \text{ to path of link 4})$$
$$V_{B/A} \qquad (\parallel \text{ to face of equivalent link 4}')$$

3. Write the acceleration equation for a_B:

$$a_B = a_A +\!\!\!\!\!\rightarrow a_{B/A}$$
$$= a^n_A +\!\!\!\!\!\rightarrow a^t_A +\!\!\!\!\!\rightarrow a^n_{B/A} +\!\!\!\!\!\rightarrow a^t_{B/A}$$

4. Determine the magnitudes and directions of the various terms.

$$a^n_A = \frac{V_A^2}{r_A} = \frac{(30)^2}{1.5} = \frac{900}{1.5} = 600 \text{ ft/sec}^2 \qquad (\parallel \text{ to link 2})$$

$$a^t_A = 0 \qquad (\omega_2 \text{ constant})$$

then $a_A = a^n_A$

$a^n_{B/A} = 0$ (no horizontal relative motion between B and A)

$a^t_{B/A}$ direction known only (*vertical*)

then $a_{B/A} = a^t_{B/A}$

5. Perform the vector addition in the acceleration diagram (Fig. 7-13e).
   a. Lay out the *direction* of a_B through the origin o.
   b. Again from the origin, lay out a_A (a^n_A).
   c. From the terminus of a_A, lay out the *direction* of $a_{B/A}$ ($a^t_{B/A}$). The intersection of this line with the a_B direction line determines a_B.

7-5. Coriolis Acceleration Component

The acceleration examples thus far have all involved two separate points on the same rigid link (with the exception of the scotch yoke). The basic acceleration relationship existing between two points on the same rigid link is expressed by the vector equation

$$a_B = a_A \dashrightarrow a_{B/A} \tag{3-20}$$

where A is the point whose motion is known and B is the point whose motion is sought. This vector equation defines the relative acceleration as between *any* two points, whether on the same rigid link or not, and whether separated or coincident. When the two points are on the same rigid link, however, the situation is usually much easier to analyze since the basic vector equation is expanded by merely breaking each of the terms into its normal and tangential components [Eq. (7-1)]. In the expanded form, the *relative-acceleration* term for the two points is expressed as

$$a_{B/A} = a^n_{B/A} \dashrightarrow a^t_{B/A} \tag{3-6}$$

When the two points involved are coincident but on different links *and* the point on one link can be thought of as tracing a path on the other link, *and* when the link containing the traced path is rotating, then an additional relative acceleration component is encountered. This additional component is known as the *Coriolis component of acceleration*. The situation just described is depicted in Fig. 7-14, where the coincident points are B_3 and B_2. As links 2 and 3 move while remaining in contact, it should be apparent that point B_3 remains a constant distance away from the surface of link 2 and can therefore be thought of as tracing a path on link 2 (extended). The radius of this path is labeled $r_{B_3/2}$, and the path rotates with link 2.

The *relative-acceleration term* for the two points in this case becomes

$$a_{B_3/B_2} = a^n_{B_3/B_2} \dashrightarrow a^t_{B_3/B_2} \dashrightarrow 2V_{B_3/B_2}\omega_2 \tag{7-2}$$

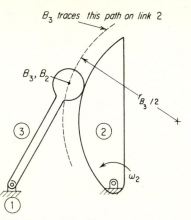

FIG. 7-14. Coincident points—one tracing path.

where the $2V\omega$ term is the Coriolis component. The terms of this expression are

$$a^n{}_{B3/B2} = \frac{(V_{B3/B2})^2}{r_{B3/2}} \qquad \text{(directed toward center of path)}$$

$$a^t{}_{B3/B2} \qquad \text{(directed tangent to path)}$$

$$V_{B3/B2} = \text{relative velocity between two points}$$

$$\omega_2 = \text{angular velocity of path (link 2)}$$

The $a^t{}_{B3/B2}$ term is always an *unknown* in this situation, since to calculate its magnitude would require knowing the angular acceleration of the motion of point B_3 relative to link 2.

One confusing aspect of the Coriolis component is its sense. In the interest of keeping confusion at a minimum, it is helpful to set forth the first of two rules.

RULE 1

The sense of the Coriolis component is the same as that of the relative-velocity vector $V_{B3/B2}$ rotated 90° in the direction of the angular velocity of the path (link associated with the path).

A second confusing aspect of the Coriolis component has to do with the ω of the $2V\omega$ component. This ω must always be the angular velocity of the link associated with the traced path. To make sure that the equation is set up consistent with this requirement, it is helpful to set forth a second rule.

RULE 2

Always write the acceleration equation as though the point doing the tracing were the unknown point.

This rule constitutes a departure from the technique utilized for two points on the same rigid link. In that case, the equation was always set up with the unknown point on the left side of the equation.

The complete general form for the acceleration equation for situations involving a Coriolis component can be developed as follows:

Since
$$a_{B_3} = a_{B_2} +\!\!\!> a_{B_3/B_2} \tag{3-20}$$

and
$$a_{B_3/B_2} = a^n{}_{B_3/B_2} +\!\!\!> a^t{}_{B_3/B_2} +\!\!\!> 2V_{B_3/B_2}\omega_2 \tag{7-2}$$

then $a^n{}_{B_3} +\!\!\!> a^t{}_{B_3} = a^n{}_{B_2} +\!\!\!> a^t{}_{B_2} +\!\!\!> a^n{}_{B_3/B_2} +\!\!\!> a^t{}_{B_3/B_2} +\!\!\!> 2V_{B_3/B_2}\omega_2$ (7-3)

This general equation, together with the two rules, comprise the format for solving acceleration problems involving coincident points. This general equation also comprises *Coriolis's law*, which may be stated as follows:

STATEMENT OF CORIOLIS'S LAW

When a point is moving along a path that has rotation, the absolute acceleration of the point is the vector sum of (*a*) the absolute acceleration of the coincident point on the path, (*b*) the acceleration of the point *relative* to the coincident point on the path, and (*c*) an additional relative-acceleration component normal to the path (the Coriolis component), whose magnitude is equal to twice the product of the velocity of the point relative to the path and the angular velocity of the path.

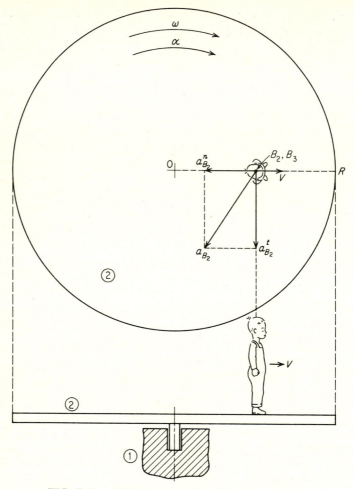

FIG. 7-15.   Coriolis acceleration illustrated.

Some feeling for the Coriolis acceleration component can be gained by visualizing a huge turntable, as shown in Fig. 7-15. If a boy B_3 stands at a point B_2 on the turntable, he feels the acceleration that point B_2 has, and he must lean in the direction of a_{B_2} to keep from falling. In other words, the acceleration of B_3 is the same as that of the coincident point B_2.

If, however, the boy starts to walk along the radial line OR toward R with a linear velocity V_{B_3/B_2}, he will feel an additional acceleration directed 90° to his path and in the direction of rotation. To compensate for this added acceleration, he must lean still further in the direction of rotation to keep from falling. This additional acceleration is caused by two factors: (*a*) the fact that his new relative linear velocity is continually being forced to change direction, and (*b*) the fact that, as he reaches each new point along his radial path, the new point has a greater tangential velocity than the point he just left (because his radius has become greater, and the linear velocity of a rotating point is proportional to its radius).

DERIVATION OF THE CORIOLIS COMPONENT

In Fig. 7-16, link 2 is rotating clockwise with a constant angular velocity ω. At the same time, link 3 is sliding along link 2 with a constant linear velocity V. After an infinitesimal period of time dt, link 2 has changed angular positions by $d\theta$, as shown. The angular displacement of link 2 alone accounts for the displacement of link 3 to point L; the linear velocity of link 3 alone accounts for the displacement of link 3 to point K. These two displacements added together would place link 3 at point M. The fact that link 3 obviously must be at N can be accounted for only by some acceleration that causes the additional displacement MN. Such an acceleration, perpendicular to the path of link 3 and in the direction of rotation, is the Coriolis acceleration component, and its magnitude can be shown to be $2V\omega$.

For circular arcs

$$d\theta = \frac{ds}{r}$$

or

$$ds = r\, d\theta$$

or

$$MN = LM\, d\theta \qquad\qquad (a)$$

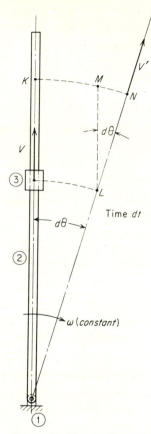

FIG. 7-16.   Proof of Coriolis acceleration.

Linear displacement may be expressed as

$$ds = \tfrac{1}{2}a(dt)^2 \qquad \text{(Art. 3-13)}$$

or

$$MN = \tfrac{1}{2}a(dt)^2 \qquad (b)$$

The linear displacement of link 3 due to the velocity V may be expressed as

$$LM = V\,dt \qquad (c)$$

The angular displacement of link 3 can be expressed as

$$d\theta = \omega \, dt \tag{d}$$

Now, substituting Eqs. (*b*) to (*d*) into Eq. (*a*) produces the expression

$$\tfrac{1}{2}a(dt)^2 = V \, dt\omega \, dt$$

or

$$a = 2V\omega$$

which is the Coriolis component of acceleration.

PATTERN OF SOLUTION FOR PROBLEMS INVOLVING CORIOLIS ACCELERATION

Since acceleration problems involving the Coriolis component are somewhat more difficult than problems involving two points on the same rigid link, it is important to make a careful diagnosis of the problem to ensure that appropriate coincident points are selected and that the acceleration equation is set up properly.

Figure 7-17 shows some examples of linkages that can be solved by the Coriolis (coincident-points) approach. The linkages shown in Fig. 7-17a, involving rotating *curved* paths, can also be worked by the *equivalent-linkage* method (these linkages are indicated by dashed lines), whereas the ones shown in Fig. 7-17b, involving rotating *straight* paths, can be worked *only* by the Coriolis approach (or by drawing motion curves, as explained in Chap. 8). In Fig. 7-17, the coincident points indicated are the logical choices since in each case it is more or less easy to visualize the path that B_2 traces on link 4. Since in each case shown, B_2 does the tracing, the acceleration equation would be set up with a_{B_2} on the left side of the equation, regardless of which link's motion is unknown.

EXAMPLE 7-6.   CAM AND FOLLOWER: CORIOLIS METHOD

In Fig. 7-18, the cam (link 2) rotates counterclockwise with a constant angular velocity of 120 rad/sec. It is required to find the angular velocity and the angular acceleration of the follower (link 3).

This problem is identical to the problem shown in Example 7-4 (Fig. 7-12) where the solution was based on the equivalent-linkage

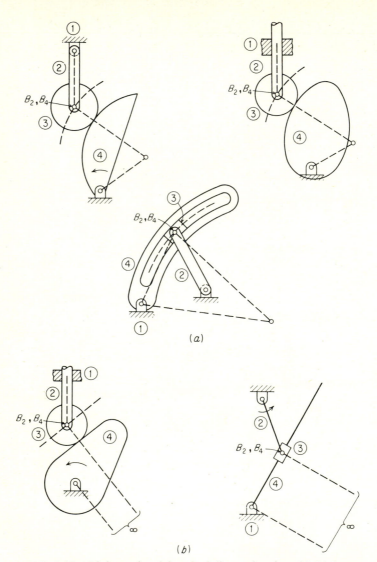

FIG. 7-17.   Linkages involving Coriolis acceleration. (*a*) Linkages involving curved rotating paths. (*b*) Linkages involving straight rotating paths.

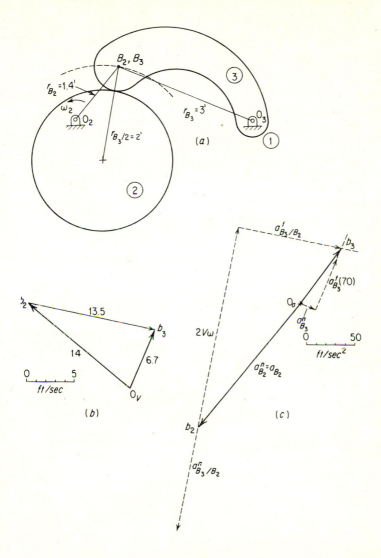

FIG. 7-18.   Circular cam mechanism: Coriolis's method. (*a*) Circular cam mechanism. (*b*) Velocity diagram. (*c*) Acceleration diagram.

concept. In this example, the solution will utilize coincident points which will result in a Coriolis component of acceleration.

SOLUTION

1. Select two coincident points, one on link 2 and one on link 3, such that one of the points traces an obvious path on the other link. Two such points are B_2 and B_3 at the center of curvature of the end of link 3. It is helpful to imagine that link 2 has an appendage that extends behind link 3 such that it overlaps point B_3.
2. Identify the path that one of the points is tracing. In this case, B_3 can be thought of as tracing a circular path on link 2, as indicated by the dashed arc; that is, as the mechanism operates, point B_3 remains a fixed distance from the surface of link 2.
3. Draw the velocity diagram, as shown in Fig. 7-18b.

$$V_{B_2} = r_{B_2}\omega_2 = (^{1.4}\!/_{12})(120) = 14 \text{ ft/sec} \qquad (\perp \text{ to } B_2O_2)$$

$$V_{B_3} \qquad (\perp \text{ to } B_3O_3)$$

$$V_{B_3/B_2} \qquad \text{(tangent to path traced by } B_3 \text{ on 2)}$$

4. Write the acceleration equation with a_{B_3} on the left side since B_3 does the tracing (Rule 2).

$$a_{B_3} = a_{B_2} \mathbin{+\!\!\!\rightarrow} a_{B_3/B_2}$$

$$a^n_{B_3} \mathbin{+\!\!\!\rightarrow} a^t_{B_3} = a^n_{B_2} \mathbin{+\!\!\!\rightarrow} a^t_{B_2} \mathbin{+\!\!\!\rightarrow} a^n_{B_3/B_2} \mathbin{+\!\!\!\rightarrow} a^t_{B_3/B_2} \mathbin{+\!\!\!\rightarrow} 2V_{B_3/B_2}\omega_2$$

5. Determine the directions and magnitudes of the terms in the above equation.

$$a^n_{B_3} = \frac{(V_{B_3})^2}{r_{B_3}} = \frac{(6.7)^2}{^3\!/_{12}} = 44.9 \text{ ft/sec}^2$$

$$(\parallel \text{ to } r_{B_3} \text{ and directed toward } O_3)$$

$$a^t_{B_3} = r_{B_3}\alpha_3 \qquad \textit{direction known only} \text{ since } \alpha_3 \text{ unknown } (\perp \text{ to } r_{B_3})$$

$$a^n_{B_2} = \frac{(V_{B_2})^2}{r_{B_2}} = \frac{(14)^2}{1.4/12} = 1{,}680 \text{ ft/sec}^2$$

$$(\parallel \text{ to } r_{B_2} \text{ and directed toward } O_2)$$

$$a^t_{B_2} = r_{B_2}\alpha_2 = 0 \quad \text{(since } \alpha_2 = 0\text{)}$$

$$\therefore a_{B_2} = a^n_{B_2}$$

$$a^n_{B_3/B_2} = \frac{(V_{B_3/B_2})^2}{r_{B_3/B_2}} = \frac{(13.5)^2}{2/12} = 1{,}093.8 \text{ ft/sec}^2$$

(directed toward center of path being traced)

$$a^t_{B_3/B_2} = r_{B_3/B_2}\alpha_{B_3/B_2}$$

direction known only (tangent to path being traced)

The α in this case is the angular acceleration of the radius vector from the center of curvature of link 2 to point B_3 and is unknown.

$$2V_{B_3/B_2}\omega_2 = (2)(13.5)(120) = 3{,}240 \text{ ft/sec}^2$$

The sense of the Coriolis component is that of V_{B_3/B_2} rotated 90° in the direction of ω_2 (ccw), or in an upward direction parallel to a line connecting the center of curvature of the path (center of link 2) to points B_3 and B_2 (Rule 1).

6. Perform the vector addition to solve the acceleration equation, as shown in Fig. 7-18c.
   a. Lay out $a^n_{B_3}$ from the origin.
   b. From the terminus of $a^n_{B_3}$, lay out the *direction* of $a^t_{B_3}$.
   c. From the origin, lay out a_{B_2} ($a_{B_2} = a^n_{B_2}$ in this case).
   d. From the terminus of a_{B_2}, lay out $a^n_{B_3/B_2}$.
   e. From the terminus of $a^n_{B_3/B_2}$, lay out the $2V\omega$ component. It has the same direction as $a^n_{B_3/B_2}$ but opposite sense, so they overlap. (Notice that $a^t_{B_3/B_2}$ was skipped, for it is an unknown.)
   f. Finally, lay out the *direction* of $a^t_{B_3/B_2}$ from the terminus of the $2V\omega$ component. The intersection of this direction line with the direction line of $a^n_{B_3}$ (step 6b) determines the senses and magnitudes of both and locates point b_3 in the acceleration diagram. The instantaneous linear acceleration of point B_3, then, is a vector drawn from the origin to point b_3.

7. Calculate the angular acceleration of link 3.

$$\alpha_3 = \frac{a^t_{B_3}}{r_{B_3}} = \frac{70}{3/12} = 280 \text{ rad/sec}^2 \quad \text{cw}$$

where the magnitude of $a^t_{B_3}$ is scaled from the acceleration diagram, and the sense of α_3 is determined by observing the sense of $a^t_{B_3}$ with respect to center O_3.

It should be noted that the results of the two methods shown in Examples 7-4 and 7-6 (Figs. 7-12 and 7-18) are identical. In fact, for this type of problem, the equivalent-linkage approach is probably more straightforward and quicker, but the application of the coincident-points approach affords an excellent verification of Coriolis's law.

EXAMPLE 7-7.   FLAT-FACED CAM: CORIOLIS METHOD

In Fig. 7-19, the cam (link 2) rotates counterclockwise with a constant angular velocity ω_2 of 18 rad/sec. It is required to find the linear acceleration a_4 of link 4.

The fact that the face of the cam is flat at the point of contact with the follower for the phase shown makes it necessary to use coincident points since it is impossible to draw an equivalent linkage (see Fig. 7-17b).

SOLUTION

1. Select the two coincident points at the center of the cam follower B_2 on link 2 and B_4 on link 4. It is more effective to think of one of the points being on link 4 rather than on link 3, since link 4 is the link of interest. Also, its motion is more obvious. Moreover, from a kinematic viewpoint, links 3 and 4 could be welded together as an integral part.
2. Identify the path involved. In this case it is easy to visualize the straight-line path that point B_4 traces on link 2, as indicated by the dashed line.
3. Draw the velocity diagram, as shown in Fig. 7-19b.

$$V_{B_2} = r_{B_2}\omega_2 = (\tfrac{4}{12})(18) = 6 \text{ ft/sec} \qquad (\perp \text{ to } B_2O_2)$$

$$V_{B_4} \qquad (\parallel \text{ to motion of link 4})$$

$$V_{B_4/B_2} \qquad (\parallel \text{ to path traced by } B_4 \text{ on 2})$$

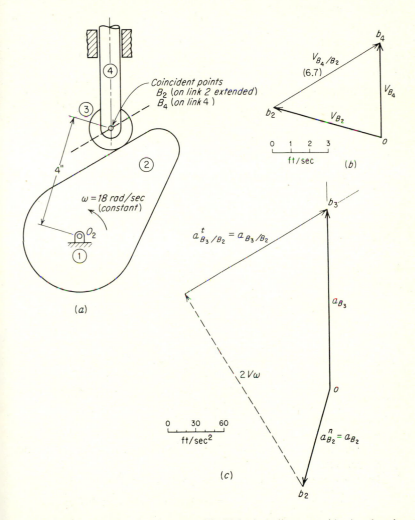

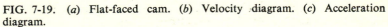

FIG. 7-19. (*a*) Flat-faced cam. (*b*) Velocity diagram. (*c*) Acceleration diagram.

4. Write the acceleration equation with a_{B_4} on the left side since B_4 does the tracing (Rule 2).

$$a_{B_4} = a_{B_2} \mathbin{+\!\!\!\!+} a_{B_4/B_2}$$
$$a_{B_4} = a^n{}_{B_2} \mathbin{+\!\!\!\!+} a^t{}_{B_2} \mathbin{+\!\!\!\!+} a^n{}_{B_4/B_2} \mathbin{+\!\!\!\!+} a^t{}_{B_4/B_2} \mathbin{+\!\!\!\!+} 2V_{B_4/B_2}\omega_2$$

5. Determine the directions and magnitudes of the terms in the above equation.

a_{B_4} *direction known only* ($\parallel$ to link 4)

$$a^n{}_{B_2} = \frac{(V_{B_2})^2}{r_{B_2}} = \frac{(6)^2}{4/12} = 108 \text{ ft/sec}^2$$

($\parallel$ to B_2O_2 directed toward O_2)

$$a^t{}_{B_2} = r_{B_2}\alpha_2 = 0 \qquad (\text{since } \alpha_2 = 0)$$

$$\therefore a_{B_2} = a^n{}_{B_2}$$

$$a^n{}_{B_4/B_2} = \frac{(V_{B_4/B_2})^2}{r_{B_4/B_2}} = 0 \qquad (\text{since } r_{B_4/B_2} = \infty)$$

$a^t{}_{B_4/B_2} = $ *direction known only*   ($\parallel$ to path traced by B_4)

$$2V_{B_4/B_2}\omega_2 = (2)(6.7)(18) = 241.2 \text{ ft/sec}^2$$

The sense of the Coriolis component is that of V_{B_4/B_2} rotated $90°$ in the direction of ω_2 (ccw), or in an upward direction perpendicular to the path traced by B_4 (Rule 1).

6. Perform the vector addition to solve the acceleration equation, as shown in Fig. 7-19c.
   a. Through the origin, lay out the *direction* of a_{B_4}. This completes the left side of the vector equation.
   b. Again from the origin, lay out a_{B_2} ($a_{B_2} = a^n{}_{B_2}$ in this case).
   c. From the terminus of a_{B_2}, lay out the $2V\omega$ component. (Notice that $a^t{}_{B_4/B_2}$ was skipped, for it is an unknown.)
   d. From the terminus of the $2V\omega$ component, lay out the direction of $a^t{}_{B_4/B_2}$. The intersection of this direction line with the direction line of a_{B_4} (step 6a) solves the vector equation and locates point b_4 in the acceleration diagram. The instantaneous linear acceleration of point B_4 (and therefore link 4) is a vector drawn from the origin to point b_4.

EXAMPLE 7-8. SHAPER MECHANISM: CORIOLIS METHOD

In Fig. 7-20, crank 2 rotates counterclockwise with a constant angular velocity ω_2 of 20 rad/sec. It is required to find the angular acceleration of link 3.

The fact that link 3 is straight makes it necessary to use coincident points since it is impossible to draw an equivalent linkage (see Fig. 7-17b). If link 3 were curved, the mechanism would be similar to the lower mechanism in Fig. 7-17a and could therefore be solved by the equivalent-linkage method.

SOLUTION

1. Select the coincident points B_3 and B_2 at the pivot point on the slider, where B_3 is on link 3 and B_2 is on link 2. Point B_2 can then be thought of as tracing a straight-line path on link 3.

2. Draw the velocity diagram, as shown in Fig. 7-20b.
$$V_{B_2} = r_{B_2}\omega_2 = (6/12)(20) = 10 \text{ ft/sec} \qquad (\perp \text{ to link 2})$$
$$V_{B_3} \qquad (\perp \text{ to link 3})$$
$$V_{B_2/B_3} \qquad (\parallel \text{ to path traced by } B_2 \text{ on 3})$$

3. Write the acceleration equation with a_{B_2} on the left side since B_2 does the tracing (Rule 2).
$$a_{B_2} = a_{B_3} \dashrightarrow a_{B_2/B_3}$$
$$a^n_{B_2} \dashrightarrow a^t_{B_2} = a^n_{B_3} \dashrightarrow a^t_{B_3} \dashrightarrow a^n_{B_2/B_3} \dashrightarrow a^t_{B_2/B_3} \dashrightarrow 2V_{B_2/B_3}\omega_3$$

4. Determine the directions and magnitudes of the terms in the above equation.
$$a^n_{B_2} = \frac{(V_{B_2})^2}{r_{B_2}} = \frac{(10)^2}{6/12} = 200 \text{ ft/sec}^2 \qquad (\parallel \text{ to link 2})$$
$$a^t_{B_2} = 0 \qquad (\text{since } \alpha_2 = 0)$$
$$\therefore a_{B_2} = a^n_{B_2}$$
$$a^n_{B_3} = \frac{(V_{B_3})^2}{r_{B_3}} = \frac{(5.1)^2}{1} = 26 \text{ ft/sec}^2 \qquad (\parallel \text{ to link 3})$$
$$a^t_{B_3} = r_{B_3}\alpha_3$$
$$\qquad \textit{direction known only}, \text{ since } \alpha_3 \text{ unknown } (\perp \text{ to link 3})$$
$$a^n_{B_2/B_3} = \frac{(V_{B_2/B_3})^2}{r_{B_2/B_3}} = 0 \qquad (\text{since } r_{B_2/B_3} = \infty)$$
$$a^t_{B_2/B_3} = r_{B_2/B_3}\alpha_{B_2/B_3} \qquad \textit{direction known only}$$

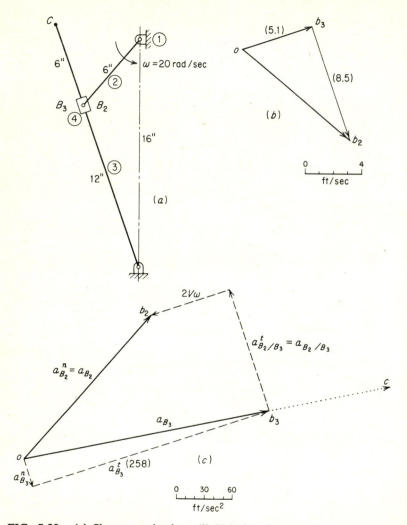

FIG. 7-20. (*a*) Shaper mechanism. (*b*) Velocity diagram. (*c*) Acceleration diagram.

where the product $r\alpha = \infty \cdot 0$ in this case, which is an indeterminate form and cannot be evaluated directly.

$$2 V_{B_2/B_3} \omega_3 = (2)(8.5)(5.1) = 86.7 \text{ ft/sec}^2$$

where $\omega_3 = \dfrac{V_{B_3}}{r_{B_3}} = \dfrac{5.1}{1} = 5.1 \text{ rad/sec}$

The sense of the Coriolis component is that of V_{B_2/B_3}, which is parallel to the path of B_2 relative to B_3 ($\searrow$) rotated 90° in the direction of ω_3 ($\curvearrowright$), or ($\swarrow$).

5. Perform the vector addition to solve the acceleration equation, as shown in Fig. 7-20c.

   a. Lay out a_{B_2} ($a_{B_2} = a^n{}_{B_2}$) from the origin. This completes the left side of the vector equation.

   b. Again, from the origin, lay out $a^n{}_{B_3}$.

   c. From the terminus of $a^n{}_{B_3}$, lay out the *direction* of $a^t{}_{B_3}$. Ordinarily an unknown vector would be skipped and left to last, but some preference should be given to keeping obvious pairs of components adjacent; that is, if a^n and a^t of point B_3 are kept adjacent, the acceleration diagram will provide the acceleration of point B_3.

   d. Since the last vector was an indefinite one, it is not possible to continue the addition process in the usual manner. It is necessary to work backward from the known sum a_{B_2}, which has already been laid out. Therefore, from the terminus of a_{B_2}, lay out the $2V\omega$ component with its terminus coinciding with the terminus of a_{B_2}.

   e. From the tail of the $2V\omega$ component, lay out the *direction* of the remaining vector $a^t{}_{B_2/B_3}$. The resulting intersection of the direction lines for $a^t{}_{B_3}$ and $a^t{}_{B_2/B_3}$ determines the magnitudes of both, and completes the acceleration diagram.

6. Calculate the angular acceleration of link 3.

$$\alpha_3 = \frac{a^t{}_{B_3}}{r_{B_3}} = \frac{258}{1} = 258 \text{ rad/sec}^2 \quad \text{cw}$$

where the magnitude of $a^t{}_{B_3}$ is scaled from the acceleration diagram, and the sense of α_3 is determined by observing the sense of $a^t{}_{B_3}$ with respect to center O_3.

7. Figure 7-20c shows the location of b_3, which provides the acceleration of point B_3 if it is desired. Also, using the acceleration-image

concept, the acceleration of point C can be obtained by the proportion $oc/ob_3 = OC/O_3B_3$.

Problems

The following problems are designed to fit on $8\frac{1}{2}$- by 11-in. paper (with the short edge horizontal). The locations for the crank 2 pivot, the origin of the velocity diagram, and the origin of the acceleration diagram are given in the form of full-scale coordinates from the left and bottom edges of the paper, respectively. Allowance is made for a $\frac{1}{4}$-in. margin at the top and right side, a $\frac{1}{2}$-in. margin on the left side, and a $\frac{3}{4}$-in. margin at the bottom.

7-1.   Crank 2 in Fig. 7-21 rotates ccw with an angular velocity of 3 rad/sec and is speeding up with an acceleration of 50 rad/sec². (a) Find the velocities of points A and B. (b) Find the angular velocities of links 3 and 4. (c) Find the accelerations of points A and B. (d) Find the angular accelerations of links 3 and 4. *Locational data:* crank 2 (2, 7); velocity diagram (7, 9); acceleration diagram (7, 2). *Scales:* space 1 in. = 1 ft; velocity 1 in. = 4 ft/sec; acceleration 1 in. = 20 ft/sec².

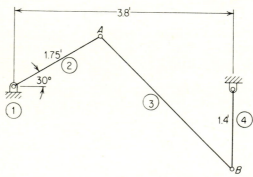

FIG. 7-21.   Prob. 7-1.

7-2.   Crank 2 in Fig. 7-22 rotates cw with an angular velocity of 60 rad/sec and is speeding up with an acceleration of 1,200 rad/sec². (a) Find the velocities of points A and B. (b) Find the angular velocity of link 3. (c) Find the accelerations of points A and B. (d) Find the angular acceleration of link 3. *Locational data:* crank 2

(3, 7); velocity diagram (6, 6); acceleration diagram (3, 6). *Scales:* space 1 in. = 1 in.; velocity 1 in. = 4 ft/sec; acceleration 1 in. = 100 ft/sec².

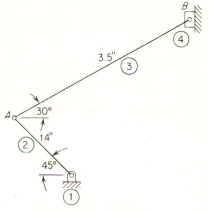

FIG. 7-22. Prob. 7-2.

7-3. Crank 2 in Fig. 7-23 rotates ccw with an angular velocity of 4 rad/sec and is slowing down with an acceleration of 5 rad/sec². (*a*) Find the velocities of points *A* and *B*. (*b*) Find the angular velocities of links 3 and 4. (*c*) Find the accelerations of points *A* and *B*. (*d*) Find the angular accelerations of links 3 and 4. *Locational data:* crank 2 (6, 3); velocity diagram (7, 9); acceleration diagram (1, 8). *Scales:* space 1 in. = 1 ft; velocity 1 in. = 2 ft/sec; acceleration 1 in. = 5 ft/sec².

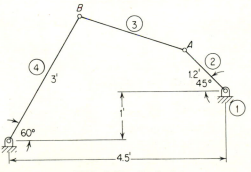

FIG. 7-23. Prob. 7-3.

7-4. Crank 2 in Fig. 7-24 rotates ccw with a constant angular velocity of 48 rad/sec. (*a*) Find the velocities of points *A* and *B*. (*b*) Find the angular velocity of link 3. (*c*) Find the accelerations of points *A* and *B*. (*d*) Find the angular acceleration of link 3. *Locational data:* crank 2 (3, 9); velocity diagram (5, 7); acceleration diagram (2, 3). *Scales:* space 1 in. = 1 in.; velocity 1 in. = 3 ft/sec; acceleration 1 in. = 100 ft/sec^2.

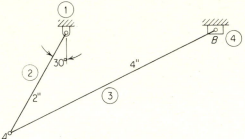

FIG. 7-24. Prob. 7-4.

7-5. Crank 2 in Fig. 7-25 rotates ccw with an angular velocity of 6 rad/sec and is slowing down with an acceleration of 10 rad/sec^2.

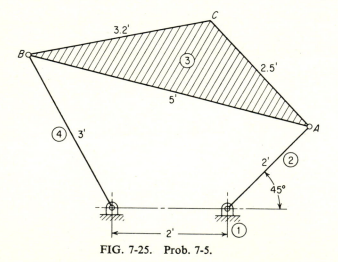

FIG. 7-25. Prob. 7-5.

(a) Find the velocities of points *A*, *B*, and *C*. (b) Find the angular velocities of links 3 and 4. (c) Find the accelerations of points *A*, *B*, and *C*. (d) Find the angular accelerations of links 3 and 4. *Locational data:* crank 2 (6, 2); velocity diagram (4, 6); acceleration diagram (6, 10). *Scales:* 1 in. = 1 ft; velocity 1 in. = 5 ft/sec; acceleration 1 in. = 20 ft/sec².

7-6. Crank 2 in Fig. 7-26 rotates cw at 191 rpm and is speeding up at the rate of 240 rad/sec². (a) Find the velocities and accelerations of points *A*, *B*, and *C*. (b) Find the angular velocity of link 3. (c) Find the angular acceleration of link 3. *Locational data:* crank 2 (2, 2); velocity diagram (5, 5); acceleration diagram (7, 10). *Scales:* space 1 in. = 1 in.; velocity 1 in. = 2 ft/sec; acceleration 1 in. = 40 ft/sec².

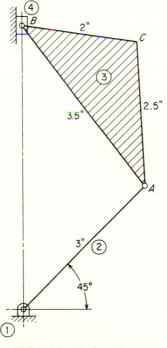

FIG. 7-26. Prob. 7-6.

7-7.   Crank 2 in Fig. 7-27 rotates ccw with an angular velocity of 12 rad/sec and is speeding up with an acceleration of 30 rad/sec². (*a*) Find the velocities of points *A*, *B*, and *C*. (*b*) Find the angular velocities of links 3 and 4. (*c*) Find the accelerations of points *A*, *B*, and *C*. (*d*) Find the angular accelerations of links 3 and 4. *Locational data:* crank 2 (3, 2½); velocity diagram (8, 5); acceleration diagram (4, 5). *Scales:* space 1 in. = 1 ft; velocity 1 in. = 10 ft/sec; acceleration 1 in. = 100 ft/sec².

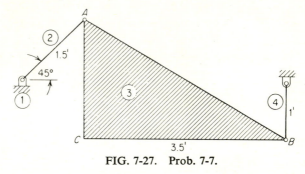

FIG. 7-27.   Prob. 7-7.

7-8.   Crank 2 in Fig. 7-28 rotates ccw with an angular velocity of 24 rad/sec and is speeding up with an acceleration of 300 rad/sec². (*a*) Find the velocities of points *A*, *B*, *C*, and *D*. (*b*) Find the angular velocities of links 3 and 4. (*c*) Find the accelerations of points *A*, *B*,

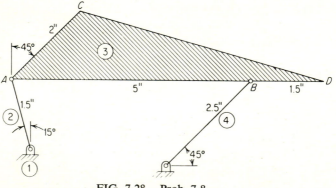

FIG. 7-28.   Prob. 7-8.

C, and *D*. (*d*) Find the angular accelerations of links 3 and 4. *Locational data:* crank 2 (1½, 1½); velocity diagram (3, 7); acceleration diagram (7, 9). *Scales:* space 1 in. = 1 in.; velocity 1 in. = 2 ft/sec; acceleration 1 in. = 20 ft/sec².

7-9.   Crank 2 in Fig. 7-29 rotates cw with an angular velocity of 6 rad/sec and is speeding up with an acceleration of 50 rad/sec². (*a*) Find the velocities of points *A*, *B*, and *C*. (*b*) Find the angular velocities of links 3 and 4. (*c*) Find the accelerations of points *A*, *B*, and *C*. (*d*) Find the angular accelerations of links 3 and 4. *Locational data:* point *A* (7½, 3); velocity diagrams (1½, 5½); acceleration diagram (4½, 10¼). *Scales:* space 1 in. = 1 ft; velocity 1 in. = 5 ft/sec; acceleration 1 in. = 30 ft/sec².

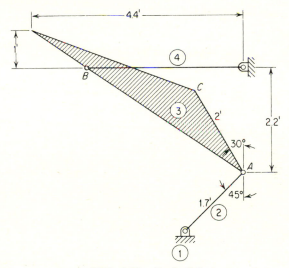

FIG. 7-29.   Prob. 7-9.

EQUIVALENT-LINKAGE METHOD

7-10.   Crank 2 in Fig. 7-30 is rotating ccw with an angular velocity of 7 rad/sec and is slowing down at the rate of 10 rad/sec². (*a*) Find the angular velocity of link 4. (*b*) Find the angular acceleration of link 4. *Locational data:* crank 2 (3, 1½); velocity diagram

$(2\frac{1}{2}, 3\frac{1}{2})$; acceleration diagram $(3\frac{1}{2}, 9\frac{1}{2})$. *Scales:* space 1 in. = 4 in.; velocity 1 in. = 3 ft/sec; acceleration 1 in. = 10 ft/sec².

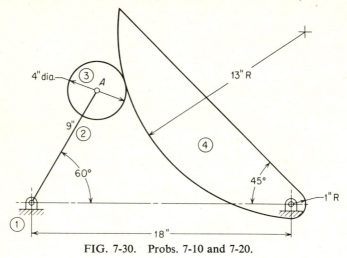

FIG. 7-30. Probs. 7-10 and 7-20.

7-11.   Link 2 in Fig. 7-31 rotates cw at the rate of 12 rad/sec and is slowing down at the rate of 60 rad/sec². (*a*) Find the angular veloc-

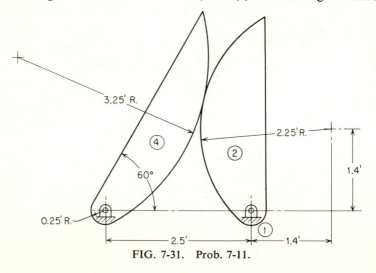

FIG. 7-31. Prob. 7-11.

ity of link 4. (*b*) Find the angular acceleration of link 4. *Locational data:* link 2 (5, 1½); velocity diagram (1, 9); acceleration diagram (6½, 10). *Scales:* space 1 in. = 1 ft; velocity 1 in. = 10 ft/sec; acceleration 1 in. = 100 ft/sec².

7-12.   The cam in Fig. 7-32 rotates cw with a constant angular velocity of 18 rad/sec. (*a*) Find the velocity of point *B* on link 4. (*b*) Find the angular velocity of link 4. (*c*) Find the acceleration of point *B* on link 4. (*d*) Find the angular acceleration of link 4. *Locational data:* link 2 (3, 2¼); velocity diagram (6½, 2½); acceleration diagram (2½, 7). *Scales:* space 1 in. = 1 in.; velocity 1 in. = 1 ft/sec; acceleration 1 in. = 10 ft/sec².

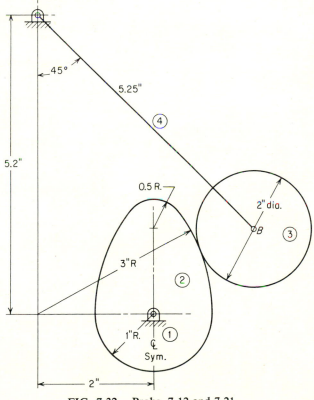

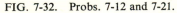

FIG. 7-32.   Probs. 7-12 and 7-21.

7-13.   Crank 2 in Fig. 7-33 rotates cw at 6 rad/sec and is speeding up at the rate of 20 rad/sec². (*a*) Find the velocity of link 4. (*b*) Find the acceleration of link 4. *Locational data:* crank 2 (2, 2); velocity diagram (1, 10); acceleration diagram (6, 10). *Scales:* space 1 in. = 6 in.; velocity 1 in. = 3 ft/sec; acceleration 1 in. = 10 ft/sec².

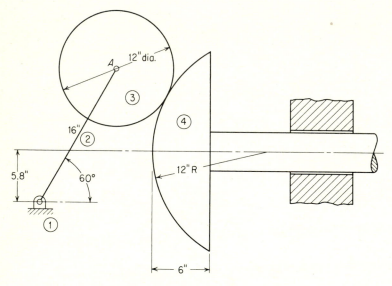

FIG. 7-33.   Prob. 7-13.

COINCIDENT POINTS AND THE CORIOLIS ACCELERATION

7-14.   The cam in Fig. 7-34 rotates ccw with a constant angular velocity of 12 rad/sec. (*a*) Find the velocity of point *B* on link 4. (*b*) Find the angular velocity of link 4. (*c*) Find the acceleration of point *B* on link 4. (*d*) Find the angular acceleration of link 4. *Locational data:* link 2 (3½, 2½); velocity diagram (3, 7); acceleration diagram (5½, 7½). *Scales:* space 1 in. = 1 in.; velocity 1 in. = 1 ft/sec; acceleration 1 in. = 30 ft/sec².

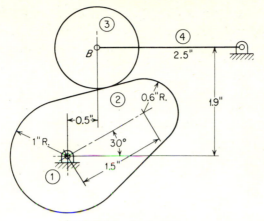

FIG. 7-34. Prob. 7-14.

7-15. Crank 2 in Fig. 7-35 rotates ccw at 110 rpm. (*a*) Find the velocities of points *A*, *B*, and *C* on link 4. (*b*) Find the angular velocity of link 4. (*c*) Find the accelerations of points *A*, *B*, and *C* on link 4. (*d*) Find the angular acceleration of link 4. *Locational data:*

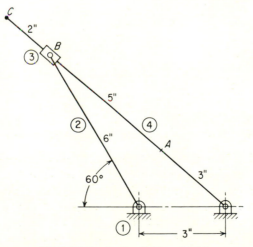

FIG. 7-35. Probs. 7-15 and 7-16.

crank 2 (5, 2); velocity diagram (3, 9); acceleration diagram (4½, 9). *Scales:* space 1 in. = 2 in.; velocity 1 in. = 2 ft/sec; acceleration 1 in. = 20 ft/sec².

7-16. Same as Prob. 7-15 except that crank 2 is speeding up at the rate of 50 rad/sec².

7-17. Crank 2 in Fig. 7-36 rotates ccw at 120 rpm. (*a*) Find the velocities of points *B* (on link 4), *C*, and *D*. (*b*) Find the angular velocities of links 4 and 5. (*c*) Find the accelerations of points *B* (on link 4), *C*, and *D*. (*d*) Find the angular accelerations of links 4 and 5. *Locational data:* crank 2 (4, 3½); velocity diagram (4, 6½); acceleration diagram (8, 9½). *Scales:* space 1 in. = 1 ft; velocity 1 in. = 4 ft/sec; acceleration 1 in. = 40 ft/sec².

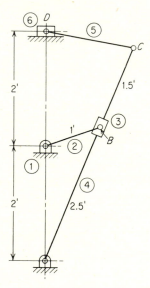

FIG. 7-36. Prob. 7-17.

7-18. Link 2 in Fig. 7-37 rotates cw at 10 rad/sec and is speeding up with an acceleration of 20 rad/sec². Find the angular acceleration of link 3. *Locational data:* link 2 (2½, 7); velocity diagram (8, 8); acceleration diagram (3, 2). *Scales:* space 1 in. = 1 ft; velocity 1 in. = 5 ft/sec; acceleration 1 in. = 30 ft/sec².

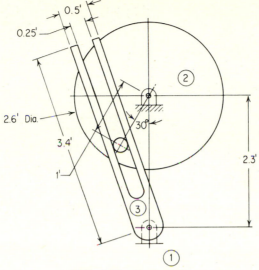

FIG. 7-37.   Prob. 7-18.

7-19.   Crank 2 in Fig. 7-38 rotates cw at 10 rad/sec and is speeding up with an acceleration of 60 rad/sec². Find the angular velocity and the angular acceleration of link 4. *Locational data:*

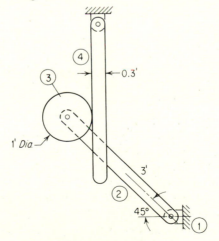

FIG. 7-38.   Prob. 7-19.

crank 2 (4, 5); velocity diagram (4½, 7); acceleration diagram (1½, 2). *Scales:* space 1 in. = 1 ft; velocity 1 in. = 10 ft/sec; acceleration 1 in. = 200 ft/sec².

7-20. Same as Prob. 7-10. Find the angular velocity and the angular acceleration of link 4 by using the coincident points A_2 and A_4. All locational data and scales remain the same.

7-21. Same as Prob. 7-12. Find the angular velocity and the angular acceleration of link 4 by using the coincident points B_2 and B_4. *Locational data:* link 2 (3, 2¼); velocity diagram (5½, 5); acceleration diagram (5, 8½). *Scales:* space 1 in. = 1 in.; velocity 1 in. = 1 ft/sec; acceleration 1 in. = 20 ft/sec².

7-22. Crank 2 in Fig. 7-39 rotates ccw at 10 rad/sec and is speeding up with an acceleration of 20 rad/sec². Find the angular velocity and the angular acceleration of link 4. *Locational data:*

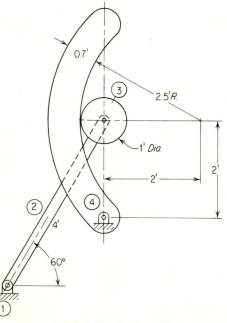

FIG. 7-39.   Prob. 7-22.

crank 2 (2, 4); velocity diagram (8, 8); acceleration diagram (5, 5). *Scales:* space 1 in. = 1 ft; velocity 1 in. = 20 ft/sec; acceleration 1 in. = 200 ft/sec².

7-23. Crank 2 in Fig. 7-40 rotates cw with a constant angular velocity of 20 rad/sec. Find the velocity and acceleration of point *C*. *Locational data:* crank 2 (2, 8); velocity diagram $(1, 4\frac{1}{2})$; acceleration diagram $(4, 5\frac{1}{2})$. *Scales:* space 1 in. = 10 in.; velocity 1 in. = 10 ft/sec; acceleration 1 in. = 400 ft/sec².

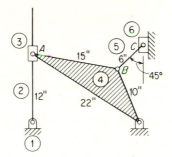

FIG. 7-40. Prob. 7-23.

motion curves

8

8-1. Introduction

In Chaps. 6 and 7, methods are presented for determining the instantaneous velocities and accelerations for the various points or links of a mechanism for a given phase. These methods, in effect, provide a snapshot of the mechanism at a particular instant. The obvious shortcoming of these methods is that it is usually impossible to tell by inspection where the *maximum* velocities or accelerations occur. Therefore, it is necessary to investigate several positions of the mechanism in an attempt to find the critical phases. It is often desirable, therefore, to have a record of the motion of a particular point or link for its entire cycle. Such a record is best presented in the form of motion curves.

8-2. Motion Curves

A complete set of motion curves for a particular point or link consists of three graphs or plots: *time-displacement*, *time-velocity*, and *time-acceleration*. The *time* in each case is usually the time required for one revolution of the driver. There are two general methods for obtaining such a set of motion curves.

The first method consists in drawing the mechanism at several phases (usually 12 or more) and constructing velocity and acceleration vector diagrams for each of these phases, using the methods presented in Chaps. 6 and 7. Then the velocities and accelerations of a particular point or link can be plotted. The *advantages* of this method are (*a*) that the accuracy is good, and (*b*) that the velocities and accelerations of *all* points in the mechanism are available from the vector diagrams. The *disadvantages* are (*a*) that there is considerable work involved in constructing velocity and acceleration diagrams for 12 or more positions of the mechanism, particularly where Coriolis accelerations are involved, and (*b*) that it is difficult to plan the locations and scales of the constructions so that the vectors will not be too small or run off the paper.

The second method consists in constructing a time-displacement curve (often referred to as a displacement diagram, especially in connection with cams) and then developing the velocity and acceleration curves by *graphical differentiation*. By this process it is possible to obtain a time-velocity curve from a time-displacement curve and a time-acceleration curve from a time-velocity curve.

Figure 8-1 shows a set of motion curves for a point in a mechanism (complete cycle not shown). These curves make it easy to see where the critical velocity and acceleration points occur. In fact, with a little practice, quite a lot of information can be inferred from the displacement curve alone. It is evident in the figure that velocity peaks occur where the displacement curve is steepest and that acceleration peaks occur where the curvature of the displacement curve is sharpest. With respect to the displacement curve, then, it is possible to make the following two statements:

STATEMENT 1

The *velocity* of a point (or link) is proportional to the *slope* of the displacement curve.

STATEMENT 2

The *acceleration* of a point (or link) is inversely proportional to the *radius of curvature* of the displacement curve.

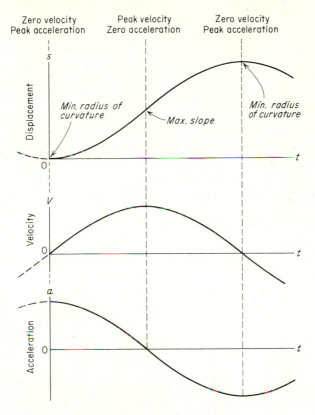

FIG. 8-1. Motion curves.

Therefore, the first step in analyzing a particular point or link in a mechanism should be to draw its displacement curve. From this curve it is possible to detect the critical phases. These critical phases can then be investigated by drawing velocity and acceleration *vector diagrams*, or the entire cycle can be investigated by developing complete motion curves by graphical differentiation.

8-3.   Graphical Differentiation

Graphical differentiation consists in obtaining the slopes of various tangents along a given curve and plotting these slope values to establish a second curve. The second curve is said to be a derivative of the first curve. The second curve can in turn be used as the basis for a third curve, which is then said to be the second derivative of the first curve.

Graphical differentiation of motion curves is based on two relationships: (*a*) the *velocity* of a point (or link) is proportional to the slope of the *displacement* curve (statement 1, Art. 8-2), and (*b*) the *acceleration* of a point (or link) is proportional to the slope of the *velocity* curve.

Figure 8-2*a* shows how the displacement of a point (in feet) varies with respect to time (in seconds). At point A line t_A is drawn tangent to the curve, and triangle LMN is drawn. The slope of the curve at point A then is MN/LM, where MN is expressed in terms of the displacement scale and LM in terms of the time scale. Since velocity is defined as $\Delta s/\Delta t$ [Eq. (3-1)], the instantaneous velocity at point A is 3.4 ft per 2 sec, or 1.7 ft/sec. This value can be plotted as A' on the velocity curve in Fig. 8-2*b*. This procedure could be repeated for other points along the displacement curve, such as points B, C, and D, to obtain the corresponding points B', C', and D' on the velocity curve. This method of constructing triangles and computing velocities is somewhat cumbersome, however, so an easier method is considered.

It should be evident that, if triangles were drawn at points B, C, and D having bases (abscissas) equal to LM, their heights would be proportional to the slopes at these points and, hence, proportional to the velocities. In other words, a velocity curve could be obtained by plotting the heights of the various triangles.

Rather than construct these triangles *along* the displacement curve, it is easier to construct *similar* triangles adjacent to the axes of the velocity curve as shown in Fig. 8-2*b*. Notice that triangle POR is similar to triangle LMN, because both are right triangles and the slope of PR has been drawn the same as the slope of LN. The size of triangle POR is arbitrary, but its size determines the location of

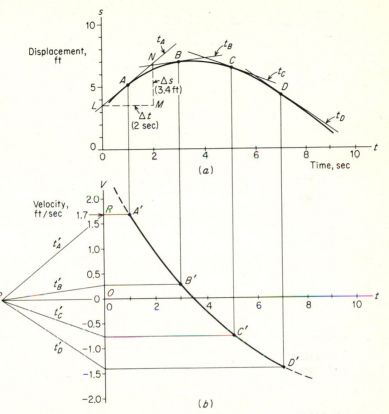

FIG. 8-2.   Graphical differentiation. (*a*) Displacement curve. (*b*) Velocity curve.

point *P*, which in turn determines the height, or *amplitude*, of the velocity curve. Since the velocity at point *A* has been computed and since it is evident that this is about the maximum velocity (steepest slope), a velocity scale can be selected that will result in a suitable height or amplitude for the velocity curve. The scale selected is laid off along the vertical axis of the velocity curve, and the known velocity at *A*, 1.7 ft/sec, is located along this scale (labeled point *R* in Fig. 8-2*b*). Line *RP* is then drawn parallel to the tangent t_A to determine the location of P, the *pole point*, and to establish PO, the *pole distance*. Point A' on the velocity curve is obtained by projecting horizontally from point R.

Now that pole point P has been established, lines t_B', t_C', and t_D' can be drawn through P parallel to the tangents t_B, t_C, and t_D at points B, C, and D, respectively. The heights of the various triangles formed by these lines determine the heights, or ordinates, of points B', C', and D' on the velocity curve.[1]

Once the velocity curve has been drawn, the process can be repeated to obtain the acceleration curve. It should be emphasized at this point, however, that, when graphical differentiation is performed twice to obtain an acceleration curve, all constructions must be made as accurately as possible, because errors introduced in either of the first two curves tend to be magnified in the acceleration curve.

EXAMPLE 8-1.   SLIDER CRANK

In Fig. 8-3a, the crank (link 2) rotates cw with a constant angular velocity of 1 rev/sec. It is required to draw the motion curves for the slider.

SOLUTION

1. Draw the mechanism for various positions of the crank (every 30° in this case) as shown in Fig. 8-3b.
2. Plot the various positions of the slider in the form of a displacement curve, as shown in Fig. 8-4a. In this case, 12 positions are plotted corresponding to every 30° of crank rotation, and the extreme right-hand position of the slider is chosen as the starting point.
3. Construct triangle LMN so that its hypotenuse is tangent to the steepest *uphill* portion of the displacement curve. The triangle may be any size but should be fairly large to facilitate measuring. It is helpful if the base (abscissa) is made equal to an integral number of units along the time scale. In this case, LM is made equal to 3

---

[1] This method and illustration were adapted from A. S. Levens, *Graphics in Engineering and Science*, pp. 475–478, John Wiley & Sons, Inc., New York, 1954.

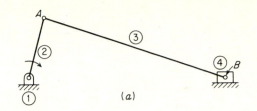

(a)

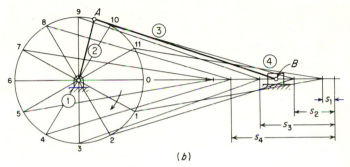

(b)

FIG. 8-3. Slider crank.

time units, or $\frac{3}{12}$ sec. The height *MN* is scaled and found to be 1.62 ft. Thus, the maximum *positive* velocity is

$$V_{\max(+)} \approx \frac{\Delta s}{\Delta t} = \frac{1.62}{\frac{3}{12}} = 6.48 \text{ ft/sec}$$

In general, if the displacement curve were not symmetrical as it is in this case, it would be desirable to construct a second triangle for the steepest *downhill* portion of the curve to determine the maximum *negative* velocity in order to obtain an idea of how far the velocity curve would dip below its horizontal axis. In this case, the positive and negative portions of the velocity curve are equal.

4. Now that the maximum positive and negative velocities are known, a velocity scale is chosen that gives a suitable height or amplitude to the velocity curve. The scale selected is laid off along the vertical axis of the velocity curve.

5. The maximum velocity of 6.48 ft/sec is located on the velocity scale (point *R* in Fig. 8-4b), and line *RP* is drawn parallel to *LN*, thus locating pole point *P*.

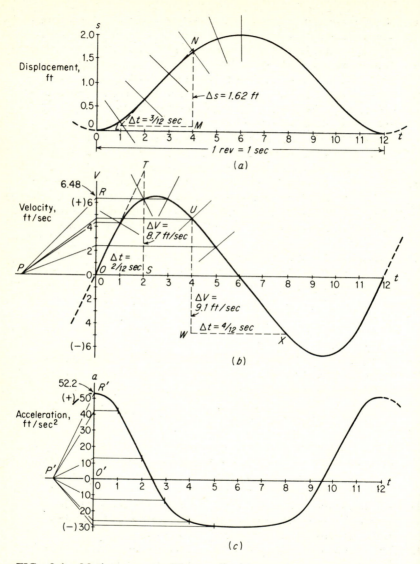

FIG. 8-4.   Motion curves: slider crank. (*a*) Displacement. (*b*) Velocity. (*c*) Acceleration.

6. Normals are next drawn along the displacement curve at the main points along the time scale. It is much easier to draw accurate normals than it is to draw accurate tangents (see Art. 4-6).

7. Through pole point *P* in Fig. 8-4*b*, draw lines perpendicular to the various normal lines that were drawn along the displacement curve. (Use the method of drawing perpendicular lines that is illustrated in Fig. 4-6.) The heights of the various triangles formed by these lines drawn through *P* determine the ordinates of the corresponding points of the velocity curve. To avoid confusion, the construction for only the top portion of the velocity curve is shown. The bottom portion is obtained in exactly the same manner.

8. The velocity curve is completed by drawing a smooth *best-fit* curve through the points just obtained. It is essential that the curve be as smooth as possible, since small dips and waves introduced by trying to force the curve through all points result in large errors in the acceleration curve.

9. Triangle *OST* is drawn to determine the maximum *positive* acceleration. Its hypotenuse is drawn tangent to the steepest *uphill* portion of the velocity curve, and its base is made equal to 2 time units, or $2/12$ sec. The height of the triangle scales 8.7 ft/sec; so

$$a_{\max(+)} \approx \frac{\Delta V}{\Delta t} = \frac{8.7}{2/12} = 52.2 \text{ ft/sec}^2$$

10. Triangle *UWX* is drawn to determine the maximum *negative* acceleration. Its base is made equal to 4 time units, or $4/12$ sec, and its height scales 9.1 ft/sec; so

$$a_{\max(-)} \approx \frac{\Delta V}{\Delta t} = \frac{9.1}{4/12} = 27.3 \text{ ft/sec}^2$$

The maximum negative acceleration is computed in this case simply to obtain an idea of how far the acceleration curve will dip below its horizontal axis. This information is helpful in planning the spacing for the acceleration curve.

11. An acceleration scale is chosen that gives a suitable height or amplitude to the acceleration curve. The scale selected is laid off along the vertical axis of the acceleration curve.

12. The maximum acceleration of 52.2 ft/sec² is now located along this vertical scale at point *R'*, and line *R'P'* is drawn parallel to line *OT*, thus locating pole point *P'*.

13.   Normals are next drawn along the velocity curve at the main points along the time scale.

14.   Through point P' in Fig. 8-4c, draw lines perpendicular to the various normal lines that were drawn along the velocity curve. The heights of the various triangles formed by these lines drawn through P' determine the ordinates of the corresponding points of the acceleration curve. The construction for only the first half of the acceleration curve is shown.

15.   The acceleration curve is completed by drawing a smooth best-fit curve through the points just obtained. In the case of the accelera-tion curve, it is not so important that the curve be drawn smooth since this curve will not be used for further differentiation.

The *mirror method* of drawing normals to a curve (Art. 4-6, Fig. 4-10) and then drawing perpendiculars to the normals (Art. 4-5, Fig. 4-6) is much more accurate than attempting to sight tangents to a curve by eye. The method is somewhat tedious, however, and can be replaced by a quicker although slightly less accurate *chord method.*

The *chord method* consists of replacing portions of the curve to be differentiated with chords as shown in the displacement diagram in Fig. 8-5a. Through the pole point P in Fig. 8-5b, draw lines parallel to the various chords that were drawn along the displacement curve. The heights of the various triangles formed by these lines determine the ordinates of the corresponding portions of the velocity curve. Notice in this case, however, that a particular ordinate for the velocity curve is plotted directly under the midpoint of the corre-sponding chord in the displacement diagram.

In the case of the slider crank in Example 8-1, the accuracy of both the velocity and acceleration curves could be improved consider-ably by plotting the velocity curve *directly* instead of deriving it from the displacement curve, that is, by obtaining the velocities of the slider at various positions by any one of the velocity methods pre-sented in Chap. 6 and plotting these values directly. The acceleration curve is then obtained by *single* rather than *double* graphical differentiation.

Figure 8-6 shows two methods for quickly obtaining the velocities of the slider at various positions. The method shown in Fig. 8-6a is

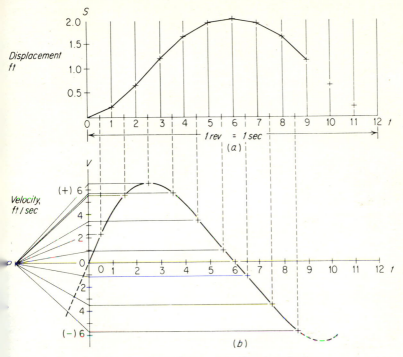

FIG. 8-5. Motion curves: chord method.

based on the *parallel-line method* (see Art. 6-5). The known velocity V_A is rotated to coincide with its radius, and a line is drawn through the terminus of the rotated vector, parallel to link 3. The point where this line intercepts the radius of B (vertical line through B) determines the magnitude of the velocity of B. It should be evident that the velocity of B at this point is horizontal and to the right, but it is more convenient to represent the velocity by a vertical vector. When the velocities of the various positions of B are obtained, these values are transferred to the velocity curve. It is sometimes useful to draw a *velocity envelope*, as shown in Fig. 8-6a.

Figure 8-6b shows the *component method* (see Art. 6-6) being used to obtain the velocities. In this method, the velocity V_A is

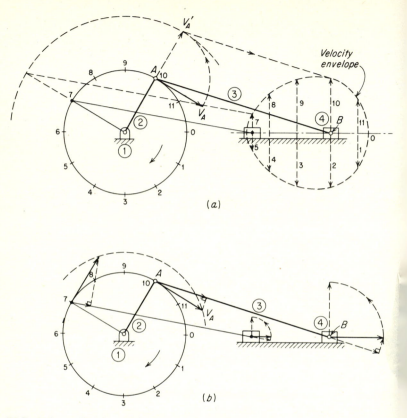

FIG. 8-6.   Velocities: slider crank. (*a*) Parallel-line method. (*b*) Component method.

resolved into two components, one parallel and one perpendicular to link 3. The parallel component is transferred to *B*, where a line drawn through its terminus and perpendicular to it determines V_B. Again, for clarity, the various velocities of *B* can be shown in the vertical position, and again, a *velocity envelope* can be drawn if desired.

It should be noted that, when the velocity curve is plotted directly, there is no need for plotting the displacement curve except to help point out the *zero* and *peak* positions of the other two curves.

EXAMPLE 8-2. FOUR-BAR MECHANISM

In Fig. 8-7a, the driver (link 2) rotates cw at 2 rev/sec. It is required to draw motion curves for the follower (link 4).

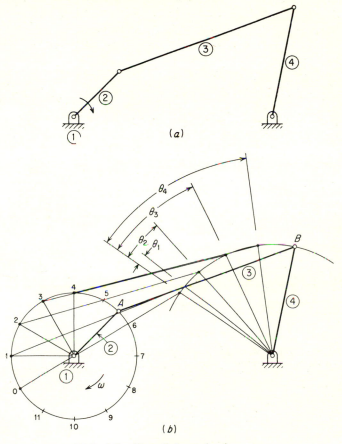

FIG. 8-7. Four-link mechanism.

SOLUTION

1. Draw the mechanism for various positions of the crank (every 30° in this case) as shown in Fig. 8-7b.

2. Measure the angular displacement of link 4 for each of the 12 positions of crank 2 (θ in Fig. 8-7*b*). The zero position of the crank is assumed to be the point at which link 4 is in its extreme left-hand position. The displacement measurements are made in degrees and then converted to radians by multiplying by 0.01745 ($360° = 2\pi$ radians, or $1° = 0.01745$ radian). The measurements and conversions are as follows:

Position of crank	θ, deg	θ, rad
0		
1	3.5	0.06
2	14.0	0.24
3	29.0	0.51
4	45.0	0.79
5	61.0	1.06
6	69.0	1.20
7	62.75	1.10
8	45.0	0.79
9	25.25	0.44
10	11.5	0.20
11	2.75	0.05
12		

3. Select suitable displacement and time scales, and plot the values of θ in the form of a displacement curve, as shown in Fig. 8-8*a*.

4. Compute the maximum *positive* angular velocity by drawing triangle *ABC* with its hypotenuse tangent to the steepest *uphill* portion of the displacement curve. With its base equal to 3 time units, or $\frac{3}{24}$ sec, its height scales 0.9 radian, or

$$\omega_{\text{max}(+)} \approx \frac{\Delta\theta}{\Delta t} = \frac{0.9}{\frac{3}{24}} = 7.2 \text{ rad/sec}$$

5. Compute the maximum *negative* angular velocity by drawing triangle *DEF* with its hypotenuse tangent to the steepest *downhill* portion of the curve. With its base equal to 2 time units, or $\frac{2}{24}$ sec, its height scales 0.73 radian, or

$$\omega_{\text{max}(-)} \approx \frac{\Delta\theta}{\Delta t} = \frac{0.73}{\frac{2}{24}} = 8.76 \text{ rad/sec}$$

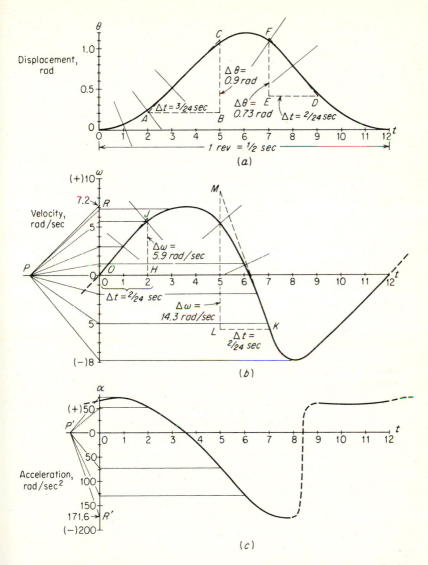

FIG. 8-8.   Motion curves: four-link mechanism. (*a*) Displacement. (*b*) Velocity. (*c*) Acceleration.

This value is needed only to plan the space required for the velocity curve.

6. A velocity scale is chosen that gives a suitable height or amplitude to the velocity curve. The scale selected is laid off along the vertical axis of the velocity curve.

7. The maximum positive angular velocity of 7.2 rad/sec is located on the velocity scale (point R in Fig. 8-8b), and line RP is drawn parallel to CA, thus locating pole point P.

8. Normals are next drawn along the displacement curve at the main points along the time scale. For purposes of this example, only five normals are drawn.

9. Through point P in Fig. 8-8b, draw lines perpendicular to the various normal lines that were drawn along the displacement curve. The heights of the various triangles formed by these lines through P determine the ordinates of the corresponding points of the velocity curve. The construction for five of the points is shown.

10. The velocity curve is completed by drawing a smooth best-fit curve through the points. A smooth curve is essential to minimize errors in the acceleration curve.

11. Triangle OHJ is drawn to determine the maximum *positive* angular acceleration.

$$\alpha_{max(+)} \approx \frac{\Delta\omega}{\Delta t} = \frac{5.9}{2/24} = 70.8 \text{ rad/sec}^2$$

12. Triangle KLM is drawn to determine the maximum *negative* angular acceleration.

$$\alpha_{max(-)} \approx \frac{\Delta\omega}{\Delta t} = \frac{14.3}{2/24} = 171.6 \text{ rad/sec}^2$$

13. An acceleration scale is chosen that gives a suitable height or amplitude to the acceleration curve. The scale selected is laid off along the vertical axis of the acceleration curve.

14. The maximum *negative* angular acceleration, 171.6 rad/sec², is located on the acceleration scale at point R', and line $R'P'$ is drawn parallel to MK, thus locating P'.

15. Normals are next drawn along the velocity curve at the main points along the time scale. For purposes of this example, only four normals are drawn.

16. Through point P' in Fig. 8-8c, draw lines perpendicular to the various normals along the velocity curve. The heights of the various

triangles formed by the lines through P' determine the ordinates of the corresponding points of the acceleration curve. The construction for four of the points is shown.

17. The acceleration curve is completed by drawing a smooth curve through the points. Notice that a portion of the acceleration curve is shown with dashed lines to indicate that this portion is somewhat indefinite because of the abrupt change.

In the preceding example, the motion curves are based on the angular motion of link 4 expressed in radians. The *angular* velocity and *angular* acceleration of this link can be obtained for any position by reading directly from the curves. The *linear* velocity and *linear* acceleration of a particular point on the link, such as B, can be obtained from

$$V_B = r\omega$$
$$a^n_B = r\omega^2$$
$$a^t_B = r\alpha$$

where ω and α are scaled directly from the motion curves and r (the radius of B) is scaled from the mechanism.

As in Example 8-1, the accuracy of both the velocity and acceleration curves in Example 8-2 can be improved considerably by plotting the velocity curve directly instead of driving it from the displacement curve. In the case of a pivoted follower, it is required to obtain the *linear* velocities of point B at various positions by one of the velocity methods presented in Chap. 6. Figure 8-9 shows two methods for obtaining quickly the velocities of B for various positions. Figure 8-9a is based on the parallel-line method (see Art. 6-5), and Fig. 8-9b is based on the component method (see Art. 6-6). Notice in Fig. 8-9a that the velocity vectors for B are shown normal to the path of B for clarity and so that a velocity envelope can be drawn if desired.

Once the linear velocities of B are found, these values can be plotted directly to obtain a velocity curve. This curve, based on *linear* velocities of B, will have the same shape as the velocity curve obtained

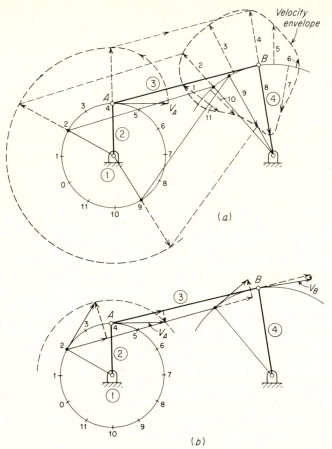

FIG. 8-9. Velocities: four-link mechanism. (*a*) Parallel-line method. (*b*) Component method.

in Fig. 8-8*b* based on *angular* velocities of link **4**. The angular velocity of link **4** for any position may be obtained by

$$\omega = \frac{V}{r}$$

where V is scaled directly from the velocity curve.

The acceleration curve for link 4 is then derived from the velocity curve by *single* graphical differentiation and is expressed in terms of the tangential acceleration of point *B*. The angular acceleration of link 4 for any position can then be obtained from

$$\alpha = \frac{a^t}{r}$$

where a^t is scaled directly from the acceleration curve.

If it were desired to have the velocity and acceleration curves expressed in terms of the angular velocity and angular acceleration of link 4 instead of the linear velocity and tangential acceleration of point *B*, the linear velocities obtained for the various positions of *B* could be converted to angular velocities of link 4 before the plotting of the velocity curve ($\omega = V/r$). A *single* graphical differentiation would then produce an acceleration curve expressed in terms of the angular acceleration of link 4.

8-4. Acceleration Discontinuities

In the examples in the preceding article, the resulting curves were relatively smooth except for the acceleration curve shown in Fig. 8-8c. In most actual cases, the displacement and velocity curves will be reasonably smooth, but it is fairly common for acceleration curves to change abruptly. In other words, although it is rare for points or links in a mechanism to have instantaneous changes in displacement or velocity, their accelerations frequently do change abruptly. For example, when the driver of an automobile quickly removes his foot from the accelerator pedal and applies it to the brake pedal, a positive acceleration is suddenly replaced by a negative acceleration. Despite this sudden change in acceleration, however, the change in the velocity of the automobile is relatively gradual.

Figure 8-10 illustrates a case in which the velocity and acceleration curves have abrupt changes. In this case, the displacement curve is deliberately constructed so that the first half of the displacement is accomplished with a constant acceleration and the second half with an equal and opposite deceleration. The equation for the displacement of a constantly accelerated particle is $s = at^2/2$, which

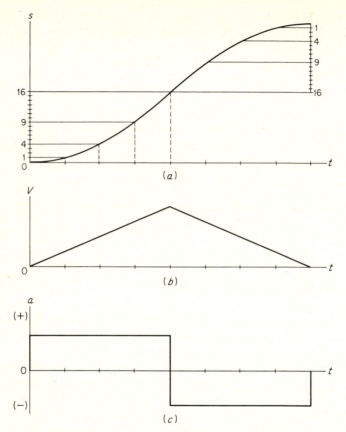

FIG. 8-10.   Discontinuities in motion curves. (*a*) Displacement.
(*b*) Velocity. (*c*) Acceleration.

shows that the displacement varies as the square of the time. The
curve in Fig. 8-10*a* is so constructed that the displacement during
equal time intervals is proportional to 1^2, 2^2, 3^2, etc., until the midpoint is reached; at that time the displacement proportions are
reversed for deceleration. From this figure it can be seen that, although the displacement curve appears innocently smooth, the
resulting velocity and acceleration curve contain abrupt changes or

discontinuities. The velocity increases and decreases, in this case, are shown to be linear (proportional to time) by the relationship $V = at$ (see Art. 3-13) for a constantly accelerated particle.

In drawing motion curves, whether by plotting points from velocity and acceleration vector diagrams or by graphical differentiation, it is important to be on the lookout for these abrupt changes. Other cases involving discontinuous velocity and acceleration curves are illustrated in Chap. 9 in connection with the motions of cam followers.

8-5. Accuracy of Graphical Differentiation

It should be emphasized that graphical differentiation is, at best, an approximate process. Its accuracy is dependent upon three main factors: (a) the number of increments into which the abscissa (time scale) is divided, (b) the care exercised in constructing tangents to the curves, and (c) the ability to fit a smooth curve to a given set of points.

The accuracy of the process increases as the number of increments along the abscissa of the displacement curve is increased. In Examples 8-1 and 8-2, the increments chosen each represented 30° of rotation of the driver. In actual practice, it is often wise to use 15° increments.

The weakest link in the whole process of graphical differentiation is the judgment factor involved in establishing accurate tangents along a curve. It requires care and practice to become consistently accurate at drawing tangents (or normals).

Finally, the ability to draw a smooth best-fit curve through a given set of points requires care and judgment as well as a good array of irregular or french curves.

A special caution is necessary regarding double graphical differentiation, where a velocity curve is derived from a displacement curve and an acceleration curve is, in turn, derived from the velocity curve, as in Examples 8-1 and 8-2. Since errors tend to be compounded, extreme care must be exercised. Fortunately, slight errors in the displacement and velocity curves, in the form of slight dips or bulges, result in exaggerated peaks in the acceleration curve. There-

fore, any rough stress computations that are based on these acceleration peaks are likely to result in a conservative design. In later design
stages, it is much easier to reduce the sizes of links than to increase
them.

8-6. Motion Curves Summarized

Motion curves provide a graphic picture of a particular point or
link for the complete cycle of the mechanism. They are useful for
pointing out critical phases, and they may be used for comparing
several proposed linkages. In general, mechanisms with the smoothest
acceleration curves will operate with the least noise, vibration, stress,
and wear.

Motion curves obtained by graphical differentiation provide
only approximate values, especially acceleration curves obtained by
double graphical differentiation. Therefore, acceleration values obtained from these curves should normally be used only for preliminary
stress computations. Those phases of a mechanism where peak
velocities or accelerations occur should be further investigated by
the more accurate methods presented in Chaps. 6 and 7.

ADVANTAGES

The advantages of motion curves may be summarized as follows:
1. They provide a good graphic picture of the motion of a particular
    point or link for its complete cycle.
2. They point out the critical phases where peak velocities or
    accelerations occur.
3. They point out abrupt *changes* in velocities and accelerations,
    which are generally undesirable.
4. They can be obtained fairly quickly by graphical differentiation,
    often eliminating the need for a time-consuming acceleration
    analysis of each individual phase of a mechanism.
5. Accelerations can be obtained for mechanisms involving coin-
    cident points on different links without resorting to the use of
    Coriolis's law.

LIMITATIONS

The disadvantages or limitations of motion curves may be summarized as follows:

1. They are approximate if obtained by graphical differentiation. An acceleration curve obtained by *double* graphical differentiation cannot be considered reliable because errors are compounded and possible discontinuities are difficult to predict.
2. A high degree of drafting skill is required to accomplish graphical differentiation satisfactorily.
3. Graphical-differentiation methods are limited to cases where the driver rotates with constant angular velocity.
4. The curves give magnitudes only. This is not usually a serious limitation, since in most cases the directions are obvious from inspection of the mechanism.
5. The motion curves portray the motion of only one particular point or link in a mechanism. If several links are to be investigated, a set of motion curves for each link must be drawn.

ANALYSIS PROCEDURE

It is difficult to set forth a definite procedure for analyzing all mechanisms, but the following general approach might be helpful:

1. Select a point or link (usually the follower) to be studied.
2. Draw the position of this point or link for a number of positions of the driver (usually a minimum of 12 equally spaced positions of the driver).
3. Plot the positions of the point or link in the form of a displacement curve, and study this curve to determine where the peak velocities and accelerations occur.
4. Two alternatives are now available:
   a. Draw velocity and acceleration *vector diagrams* for the critical positions only.
   b. Draw complete velocity and acceleration *curves*. Again there are alternatives, this time three:

(1) Determine the velocities and accelerations for the various positions of the point or link by the vector methods presented in Chaps. 6 and 7, and plot these values in the form of velocity and acceleration curves.

(2) Obtain the *velocities only* for the various positions of the point or link by one of the vector methods in Chap. 6, and plot these velocities in the form of a velocity curve from which an acceleration curve is obtained by *single* graphical differentiation.

(3) From the displacement curve of the point or link (item 3), obtain the velocity and acceleration curves by *double* graphical differentiation.

If it is desired to draw motion curves for a point that does not follow a straight-line or circular path, it may be necessary to make two displacement curves: one for the x component of its displacement, the other for the y component of its displacement. Both of these curves must then be graphically differentiated to obtain the x and y components of its velocities and accelerations at its various phases.

Problems

PROBLEM LAYOUT INSTRUCTIONS

It is recommended that the mechanism and motion curves be laid out on two separate $8\frac{1}{2}$- by 11-in. sheets placed side by side with their short dimensions parallel to the T square. (Both could be laid out on one 11- by 17-in. sheet if preferred.) The curves should be laid out close to the right side of the paper to allow space for the pole distances on the left side. It is further recommended that the abscissas (time scales) of the curves be drawn 6 in. long and that the displacement curve be approximately 2 in. high. With careful selection of velocity and acceleration scales, all three curves should fit on the same sheet.

8-1. Crank 2 in Fig. 8-11 rotates cw with a constant angular velocity of 1 rev/sec. Draw the motion curves for the slider. The beginning of the curves should correspond to the extreme left position of the slider.

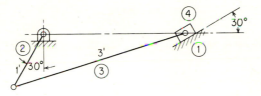

FIG. 8-11. Prob. 8-1.

8-2. Crank 2 in Fig. 8-12 rotates ccw with a constant angular velocity of 3 rev/sec. Draw the motion curves for the slider. The beginning of the curves should correspond to the extreme left position of the slider.

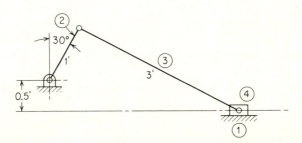

FIG. 8-12. Prob. 8-2.

8-3. Crank 2 in Fig. 8-13 rotates cw at 120 rpm. Draw the motion curves for link 4. The beginning of the curves should correspond to the extreme left position of link 4.

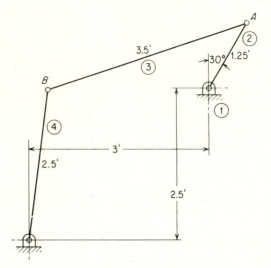

FIG. 8-13. Probs. 8-3 and 8-4.

8-4. Same as preceding problem, but draw the motion curves for point *B*.

8-5. Crank 2 in Fig. 8-14 rotates cw at 72 rpm. Draw the motion curves for link 4. The beginning of the motion curves should correspond to the extreme left position of link 4.

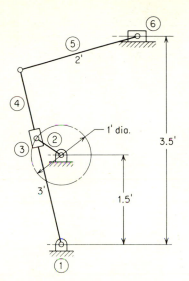

FIG. 8-14.   Probs. 8-5 and 8-6.

8-6.   Same as preceding problem, but draw the motion curves for the slider (link 6).

cams

9

9-1. Introduction

Cams are very important and frequently occurring elements in many types of machines—especially automatic machines. They provide a simple means of obtaining unusual or irregular motions that would be difficult if not impossible to obtain with other types of linkages. Cams are the heart of such automatic devices as automatic machine tools, record changers, mechanical calculators, cash registers, and many other devices.

   A cam is a link having an irregular surface or groove that imparts motion to a follower, which slides or rolls over its surface or in its groove. The simplest cam mechanism consists of a cam, a follower, and a frame. In rare instances the cam motion is *inverted*, and the cam becomes the follower. An example of this is the mechanical push-button tuner on some radios.

9-2. Cam Types

Cams are usually in the form of disks or cylinders, but occasionally they take other forms, such as translating plates. Figure 9-1 shows several types of cams. Most cam followers, particularly on radial cams, are held against the cam profile by springs or by gravity, but occasionally cam followers are positive in action, as shown in Fig.

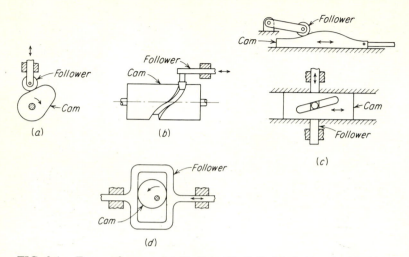

FIG. 9-1.   Types of cams. (*a*) Radial. (*b*) Cylindrical (drum). (*c*) Linear. (*d*) Positive-action.

9-1*d*. The discussion and explanation in this chapter will be confined to radial cams, because (*a*) these are by far the most frequently occurring cams, (*b*) the theory of cams can be more clearly presented with reference to this type of cam, and (*c*) once the theory of cams is understood, it is fairly easy to extend the methods to any type of cam.

9-3.   Form and Path of Follower

The actual form of the follower of a radial cam may vary considerably. It may consist of a point or knife edge (rarely used), a rounded or curved surface, a roller, or a flat surface. These various forms are illustrated in Fig. 9-2*a*. The path of the follower may also vary. It can be along a straight line that passes through the center of rotation of the cam, as shown in Fig. 9-2*a*, along a straight line that is offset from the center of rotation of the cam, as shown in Fig. 9-2*b*, or along a curved path such as an arc, as shown in Fig. 9-2*c*.

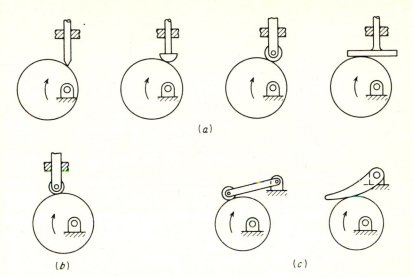

FIG. 9-2. Forms and paths of cam followers. (*a*) In-line. (*b*) Offset. (*c*) Pivoted.

9-4. Displacement Diagrams

The motion of the follower is of primary interest in the analysis of existing cams or in the design of new cams. Since cams vary widely in size and form and the motions of followers vary widely, it is easier to analyze the motion of cam followers if their motion is plotted as a graph, where the displacement of the follower is laid off parallel to the *y* axis and the angular displacement of the cam (one revolution) is laid off along the *x* axis. The resulting graph, shown in Fig. 9-3, is actually a *time-displacement curve*; when used in connection with cams, it is more often referred to as a *displacement diagram*. The displacement of the follower is usually laid out full size, but the time scale (one revolution of the cam) may be laid out to any convenient scale. Points are usually plotted for every 15° or 30° of cam rotation, and the points are connected with a smooth curve.

Since displacement diagrams are actually time-displacement curves, they provide a good graphic picture of the cam-follower

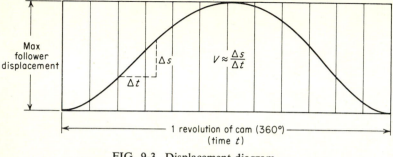

FIG. 9-3. Displacement diagram.

motion. As explained in Chap. 8, by inspection of this curve it is possible to detect the critical phases of the cam follower, that is, those phases at which peak velocities or accelerations occur (the velocity is proportional to the slope, and the acceleration is inversely proportional to the radius of curvature). In fact, the displacement diagram can serve as the basis for developing complete velocity and acceleration curves by methods presented in Chap. 8.

Displacement diagrams are indispensable not only in analyzing existing cams but also in designing new cams. They make it possible to lay out the desired follower motion linearly with no regard for the physical cam-and-follower arrangements. This greatly simplifies the geometric constructions and analysis procedures.

9-5. Drawing Displacement Diagram from Cam Profile

Figure 9-4 shows how a displacement diagram can be developed from the cam profile. In this case, the cam rotates ccw, as shown by the arrow. For purposes of analysis, the same effect is obtained if the cam is thought of as stationary and the follower as rotating cw. At 30°, the follower is drawn in tangent to the cam profile, and the amount of displacement at this position is obtained by drawing an arc from its center around to the 0° line and noting the amount of rise from the initial position. This displacement is then plotted on the 30° line in the displacement diagram. After this is done for every

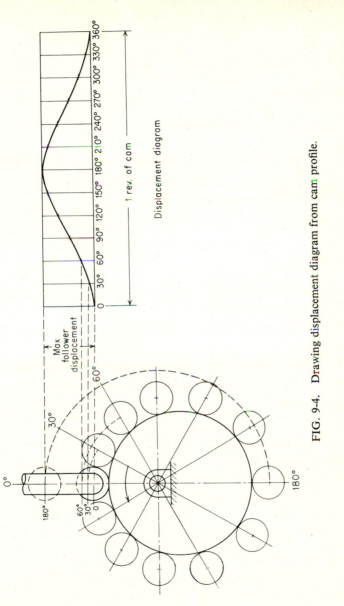

FIG. 9-4. Drawing displacement diagram from cam profile.

30°, the curve can be drawn in as shown. The technique of drawing a displacement diagram for an existing cam varies somewhat for different follower shapes and arrangements. These variations in technique will be illustrated in later examples.

9-6.   Motions Used for Cam Followers

In designing cams, the chief interest is the motion of the follower; the cam profile is merely a means of obtaining this motion. In fact, for different shapes and arrangements of followers, different cam profiles are required to obtain the same motion.

The designer may choose any type of motion he wants—the possibilities are infinite. For unimportant slow-moving cams, the cam profile often consists of combinations of arcs and straight lines, as shown in Fig. 9-5. Crude cams of this type, often referred to as *tangent cams*, are easy to draw and easy to manufacture.

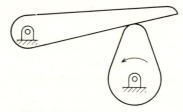

FIG. 9-5.   Tangent cam.

For high-speed applications, it is not enough merely to provide a given displacement at a given instant. It becomes necessary to attain the required displacement with a minimum of noise, vibration, and stresses. If the cam in Fig. 9-5, for example, were operated at high speed, the cam and follower would be subjected to very high stresses or the follower would tend to *float*, that is, fail to remain in contact with the cam during the entire cycle.

The motions that are most commonly used are:

1.   Uniform-velocity (straight-line) motion
2.   Simple harmonic motion (SHM)

3. Uniformly accelerated (parabolic) motion
4. Modified uniform-velocity motion
   a. Arc method
   b. Uniform-acceleration method
5. Cycloidal motion

UNIFORM-VELOCITY (STRAIGHT-LINE) MOTION

If the follower is to move with uniform velocity, its displacement must be the same for equal units of time (equal angles of cam rotation), as shown in Fig. 9-6. Its *curve* in the displacement diagram,

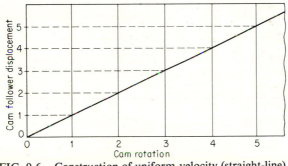

FIG. 9-6.   Construction of uniform-velocity (straight-line) motion.

therefore, is a straight line. This motion is seldom used in its pure form except for nonprecision low-speed applications because of the theoretically infinite accelerations[1] that occur at the beginning and end of the displacement. The actual motion of the follower is not always so abrupt as the diagram would indicate because these sharp points are usually rounded slightly on the actual cam profile.

EXAMPLE 9-1.   UNIFORM-VELOCITY MOTION

Lay out the displacement diagram for a cam follower that is to have the following motion:

[1] The accelerations are infinitely large in theory, but actually the members deform or separate, which reduces the accelerations involved.

Dwell 30° (at rest)
Rise 2 in. in 90° (uniform velocity)
Dwell 30°
Fall 2 in. in 60° (uniform velocity)
Dwell 150°

SOLUTION

The various displacements in the displacement diagram consist merely of straight lines, as shown in Fig. 9-7.

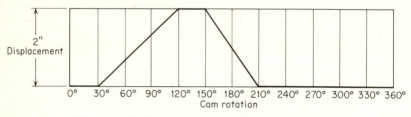

FIG. 9-7.   Uniform-velocity (straight-line) motion.

SIMPLE HARMONIC MOTION (SHM)

As a point moves around the circumference of a circle with a constant velocity, its projection on the diameter of the circle moves with simple harmonic motion. In Fig. 9-8a, as point *P* moves around

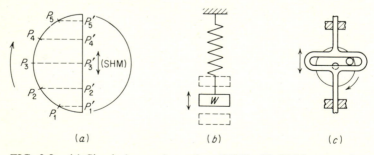

FIG. 9-8.   (*a*) Simple harmonic motion (SHM). (*b*) Weight on spring. (*c*) Scotch yoke.

the circle from P_1 to P_2, etc., its projection P'_1, P'_2, etc., moves with harmonic motion. This relationship is the basis for the construction shown in Fig. 9-9. In this construction it should be noted that the diameter of the semicircle is equal to the follower rise and that the number of divisions around the semicircle agrees with the number of divisions along the time axis (angle of cam rotation).

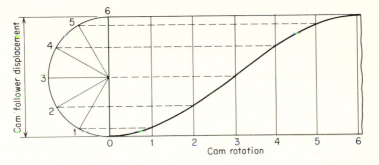

FIG. 9-9. Construction of simple harmonic motion (SHM).

Figure 9-8b shows that a weight on a spring also generates simple harmonic motion; this is why this type of motion is often referred to as *vibratory* motion. Figure 9-8c shows the scotch yoke, which also generates simple harmonic motion.

EXAMPLE 9-2. SIMPLE HARMONIC MOTION

Lay out the displacement diagram for a cam follower that is to have the following motion:

> Rise 2 in. in 120° (SHM)
> Dwell 30°
> Fall 1 in. in 90° (SHM)
> Dwell 30°
> Fall 1 in. in 60° (SHM)
> Dwell 30°

SOLUTION

For each rise or fall it is necessary to draw a semicircle, as shown in Fig. 9-10. Note that the number of divisions of each semicircle must agree with the number of horizontal divisions involved.

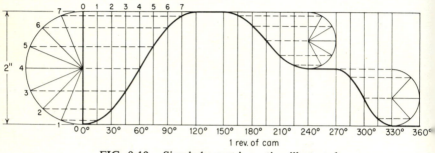

FIG. 9-10. Simple harmonic motion illustrated.

UNIFORMLY ACCELERATED (PARABOLIC) MOTION[1]

A motion of uniform or constant acceleration, often called *parabolic motion* because a graph of its equation ($s = at^2/2$) is a parabola, is the motion of a freely falling body. It is motion in which the displacement taking place in each successive interval of time is proportional to the square of the time. Figure 9-11 shows how the displacement s of a falling object compares with arbitrary time units.

In laying out this type of motion in a displacement diagram, a given displacement is divided into two halves—the first half is uniformly accelerated, and the second half is uniformly decelerated. It is, therefore, necessary that the horizontal distance involved in the total displacement be divided into an *even* number of divisions. Figure 9-12 illustrates the basic construction for a 2-in. rise in 120°, where the 120° has been divided into eight equal divisions. If the 120° had been divided into 10 equal divisions, the *half-displacement* of 1 in.

[1] This type of motion is often referred to as *uniformly accelerated and retarded motion* (UARM).

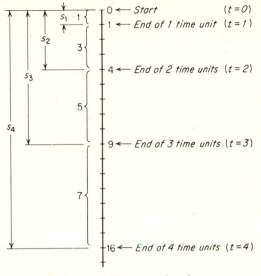

FIG. 9-11.   Uniform acceleration.

would require 25 divisions (5^2). Notice in the figure that the deceleration portion of the curve is just the reverse of the acceleration portion and that points for it can be transferred from the first part of the curve.

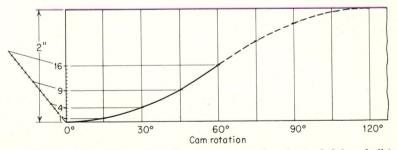

FIG. 9-12.   Construction of uniformly accelerated and retarded (parabolic) motion.

Figure 9-13 shows an alternative method of constructing this type of motion. In this case, the *half-displacement* is divided into equal-sized divisions corresponding to the number of horizontal divisions.

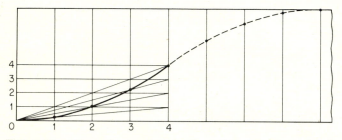

FIG. 9-13.   Alternative construction for uniformly accelerated and retarded motion.

EXAMPLE 9-3.   UNIFORM ACCELERATION

Lay out the displacement diagram for a cam follower that is to have the following motion:

> Rise 2 in. in 120° (uniform acceleration)
> Dwell 30°
> Fall 1 in. in 90° (uniform acceleration)
> Dwell 30°
> Fall 1 in. in 90° (uniform acceleration)

SOLUTION

Figure 9-14 shows the displacement diagram for the above motion. The two 1-in. displacements are identical in shape, since they take place in the same amount of cam rotation. Therefore, one construction serves both.

MODIFIED UNIFORM-VELOCITY MOTION

As mentioned previously, uniform-velocity motion is seldom used in its pure form because the abrupt starts and stops cause theoretically infinite accelerations. This type of motion is usually

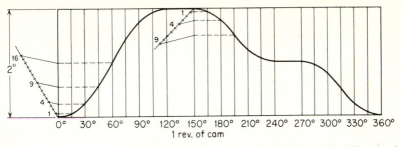

FIG. 9-14. Uniformly accelerated and retarded (parabolic) motion illustrated.

modified by introducing a short curve at each end of the straight-line portion. The two methods used are (a) to introduce an arc at each end, and (b) to introduce a period of uniform acceleration (or deceleration) at each end.

The *arc method* consists merely in introducing arcs at the beginning and end of the displacement period. The size of the arcs is arbitrary, but they are usually drawn with a radius equal to one-half the displacement. The arcs are drawn first to an indefinite length; then a straight line is drawn tangent to both arcs. Notice in Fig. 9-15 that the resulting straight-line portion of the curve is steeper than the original curve. This means that the velocity of the follower is increased, but the follower now starts and stops gradually.

The *uniform-acceleration method* consists in introducing short periods of uniform acceleration or deceleration at both ends of the

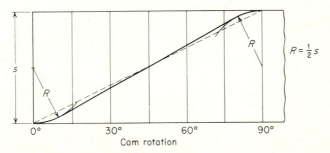

FIG. 9-15. Modified uniform-velocity motion: arc method.

displacement period. The first step involved is to decide arbitrarily on the proportion of these periods to the total period of the displacement. For example, in Fig. 9-16*a*, the 90° period is broken into 30° of acceleration, 30° of uniform velocity, and 30° of deceleration. The actual construction then consists in drawing the diagonal line *BE*, where *B* is in the middle of the acceleration period and *E* is in

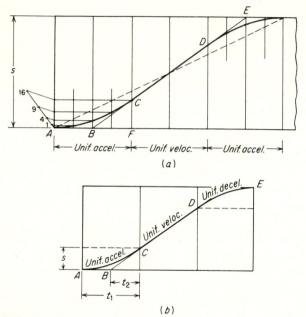

FIG. 9-16.   Modified uniform-velocity motion: uniform-acceleration method.

the middle of the deceleration period. This line *BE* locates points *C* and *D*, which determine the vertical portions of the displacement in which acceleration takes place. The construction for these two portions of the curve is then the same as the construction shown in Fig. 9-12 or 9-13.

It is important to note that, once the displacement period has been arbitrarily divided into acceleration, uniform-velocity, and deceleration periods, there is only one possible slope for the line *BE* that will make this line tangent to the acceleration and deceleration curves at the precise points where the displacement period has been divided. A line drawn between the midpoints of the acceleration and deceleration periods (*B* and *E*) is the only line that will satisfy this condition.

PROOF THAT LINE DRAWN BETWEEN MIDPOINTS OF ACCELERATION PERIODS IS THE CORRECT LINE

In Fig. 9-16*b*, consider two points starting from *A* and *B*, respectively. Point *A* starts from rest and accelerates uniformly until reaching *C*. Point *B* moves with uniform velocity from *B* to *C*. If both points are to have the same velocity (slope) at *C*, then point *A* will require twice as much time to complete its journey as point *B* will, because *A* starts from rest. This relationship can be shown algebraically as follows:

Curve *AC* represents a point starting from rest and accelerating uniformly:

$$s = \frac{at_1^2}{2} \tag{1}$$

Its final velocity at *C* is

$$V = at_1 \tag{2}$$

Line *BC* represents a point moving with uniform velocity, so

$$s = Vt_2 \tag{3}$$

Since both points move through the same distance *s*, Eqs. (1) and (3) may be equated to each other:

$$\frac{at_1^2}{2} = Vt_2 \tag{4}$$

Also, since the velocities of the two points must be equal at *C*, the value for *V* in Eq. (2) may be substituted into Eq. (4).

$$\frac{at_1^2}{2} = (at_1)t_2 \tag{5}$$

Then

$$t_2 = \frac{t_1}{2}$$

CYCLOIDAL MOTION

If a circle rolls along a straight line without slipping, a point on its circumference traces a curve that is called a *cycloid*. Figure 9-17 shows how such motion is laid out in a displacement diagram. Line *AB* is drawn and is extended to some point such as *C*. A circle is drawn at *C* whose circumference is equal to the displacement *s* or

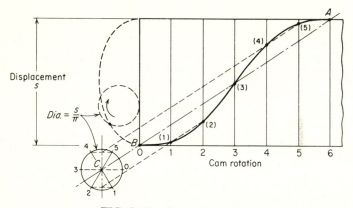

FIG. 9-17. Cycloidal motion.

whose diameter is equal to s/π. The circumference of this circle is divided into a number of parts corresponding to the number of divisions along the horizontal scale. The points around the circle are projected to the vertical center line of the circle and then parallel to line *AB* to the corresponding vertical lines in the displacement diagram.

9-7. Comparison of Cam-follower Motions

Needless to say, there is no one best motion for all cam followers. Many factors influence the selection of a follower motion for a particular situation, such as size, operating speeds, function, materials, cost, life expectancy, permissible noise and vibration, and available fabrication facilities.

Figure 9-18 shows the motion curves for the various motions described in this chapter. Figure 9-18a makes it obvious why *uniform-velocity motion* is undesirable, where high speeds are involved, because of the infinite accelerations.

Figure 9-18b shows that the acceleration curve for *simple harmonic motion* is smooth only if the rise and fall periods are both 180°. If these periods are unequal or are adjacent to dwell periods, then discontinuities appear in the acceleration curve.

Figure 9-18c shows that *uniformly accelerated (parabolic) motion* also results in a severely discontinuous acceleration curve and so unsuitable for high-speed applications. The chief advantage of this type of motion is that a given displacement in a given length of time is accomplished with the least acceleration of any motion.

Figure 9-18d shows the characteristics of *modified uniform-velocity motion*. Although the accelerations have been reduced to finite values as compared with the *unmodified* uniform-velocity motion, there still remain severe discontinuities which limit the usefulness of this motion.

Figure 9-18e makes it clear why *cycloidal motion* is considered ideal. One cycloidal displacement blends smoothly into any other cycloidal displacement. The acceleration curve is abrupt only if the curve is adjacent to a dwell period, and even then the abruptness is not so severe as it would be for other types of motion.

From the above observations it appears that either cycloidal motion or simple harmonic motion where the rise and fall periods are both equal to 180° provides the best possible cam-follower motion for high-speed applications where vibration, noise, and stresses are important factors. The other motion types must necessarily be restricted to low-speed applications.

9-8. Construction of the Cam Profile

Once the desired follower motion has been described with a displacement diagram, it is necessary to construct an actual cam profile that will accomplish this motion. The exact cam profile will vary depending upon the size of the cam and upon the size, shape, and path of

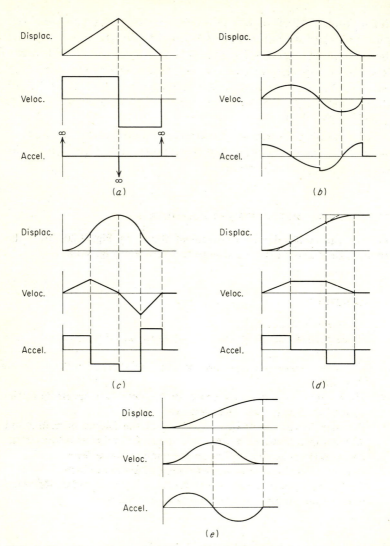

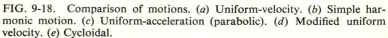

FIG. 9-18. Comparison of motions. (*a*) Uniform-velocity. (*b*) Simple harmonic motion. (*c*) Uniform-acceleration (parabolic). (*d*) Modified uniform velocity. (*e*) Cycloidal.

the follower. Therefore, it is necessary to decide on the physical size and arrangement of the cam and follower before the cam profile can be constructed. Figure 9-2 shows some of the possible arrangements. The rest of this chapter will consist of example problems in which cam profiles are generated for the most common arrangements. All the examples will be based on the same follower motion, which is shown in Fig. 9-19.

To simplify the layout procedure, it is convenient to think of the cam as being stationary and the follower as rotating about the cam in the opposite direction.

EXAMPLE 9-4. IN-LINE ROLLER FOLLOWER

For the cam-and-follower arrangement shown in Fig. 9-20, draw the cam profile based on the displacement diagram in Fig. 9-19.

SOLUTION

1. Draw the *base circle*.
2. Draw the follower in its *home* position (0° position), tangent to the base circle.
3. Draw the *reference circle* through the center of the follower in its 0° position.
4. Draw radial lines from the center of the cam, corresponding to the vertical lines in the displacement diagram.
5. Transfer displacements s_1, s_2, s_3, etc., from the displacement diagram to the appropriate radial lines, measuring from the reference circle.
6. Draw in the follower outline on the various radial lines.
7. Draw a smooth curve tangent to these follower outlines.
   To draw a smooth curve, it may be necessary to transfer additional intermediate points from the displacement diagram.

EXAMPLE 9-5. OFFSET ROLLER FOLLOWER

For the cam-and-follower arrangement shown in Fig. 9-21, draw the cam profile based on the displacement diagram in Fig. 9-19.

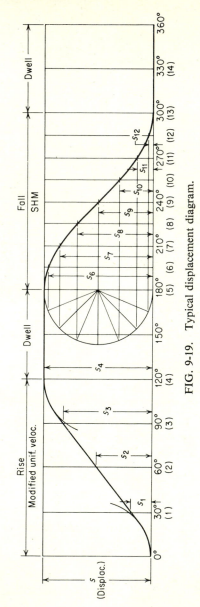

FIG. 9-19.   Typical displacement diagram.

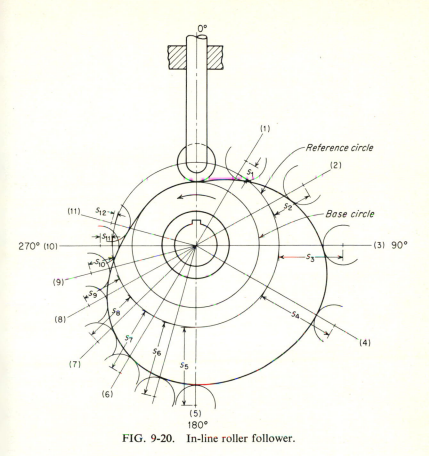

FIG. 9-20. In-line roller follower.

SOLUTION

1. Draw the *base circle*.
2. Draw the follower in its *home* position (0° position) tangent to the base circle.
3. Draw the *reference circle* through the center of the follower in its home position.
4. Draw the *offset circle* tangent to the follower center line.

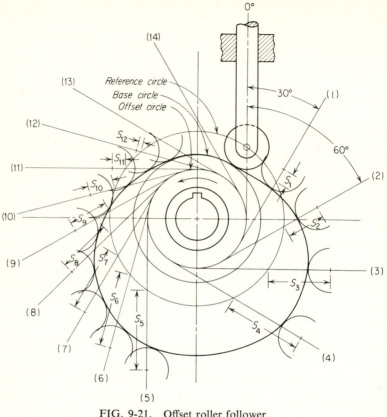

FIG. 9-21.   Offset roller follower.

5. Divide the offset circle into a number of divisions corresponding to the divisions in the displacement diagram, and number accordingly.
6. Draw tangents to the offset circle at each number.
7. Lay off the various displacements s_1, s_2, s_3, etc., along the appropriate tangent line, measuring from the reference circle.
8. Draw in the follower outlines on the various tangent lines.
9. Draw a smooth curve tangent to these follower outlines.

Example 9-6. Pivoted Roller Follower

For the cam-and-follower arrangement shown in Fig. 9-22, draw the cam profile based on the displacement diagram in Fig. 9-19.

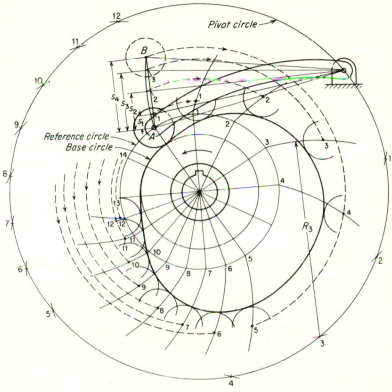

FIG. 9-22. Pivoted roller follower.

SOLUTION

1. Draw the *base circle*.
2. Draw the follower in its *home* position, tangent to the base circle.
3. Draw the *reference circle* through the center of the follower.

4. Locate points around the reference circle corresponding to the divisions in the displacement diagram, and number accordingly.
5. Draw a *pivot circle* through the follower pivot.
6. Locate the pivot points around the pivot circle corresponding to each point on the reference circle, and number accordingly.
7. From each of the pivot points, draw an arc whose radius is equal to the length of the follower arm.
8. At the zero position, draw the two extreme positions of the follower lever by laying off the chord *AB* equal to the maximum displacement.
9. Lay off the various displacements s_1, s_2, s_3, etc., along this chord; then project these points to the arc *AB*.
10. Rotate each of the points on arc *AB* to its proper position around the cam profile.
11. Draw in the follower outline at each of the points just located.
12. Draw a smooth curve tangent to the follower outlines.

In the last example, the displacement diagram was assumed to be based on the chordal distance *AB* in Fig. 9-22. The error introduced by using the chordal distance rather than the arc distance is usually disregarded. This error increases, however, as the radius of the follower decreases. If the error is felt to be too large, it can be eliminated by laying out the displacement diagram in terms of the *angular* displacement of the follower and transferring *angles* rather than *chordal distances* to the cam profile. The angular displacements may be expressed either in degrees or in radians, but, if the displacement curve is to be used subsequently as a basis for obtaining velocities and accelerations of the follower, its vertical scale should be converted to radian measure.

EXAMPLE 9-7. FLAT-FACED FOLLOWER

For the cam-and-follower arrangement shown in Fig. 9-23, draw the cam profile based on the displacement diagram in Fig. 9-19.

SOLUTION

1. Draw the base circle, which in this case also serves as the reference circle.
2. Draw the follower in its *home* position, tangent to the base circle.

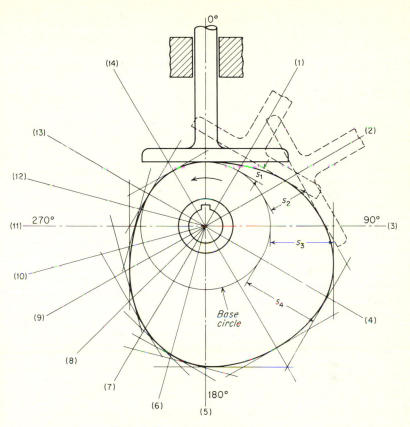

FIG. 9-23. Flat-faced follower.

3. Draw radial lines corresponding to the divisions in the displacement diagram, and number accordingly.
4. Draw in the follower outline on the various radial lines by laying off the appropriate displacements and drawing lines perpendicular to the radial lines.
5. Draw a smooth curve tangent to the follower outlines.
    Notice in the figure that the tangent points usually fall at the mid-points of the inner sides of the small triangles that are formed around the periphery of the cam.

The preceding examples show layout procedures for some of the more common cam-and-follower arrangements. There are countless possible arrangements for cams and followers, and there are countless layout procedures. Therefore, these examples should be regarded as general rather than specific guides.

9-9.   Pressure Angle

The angle between the path of the follower and the common normal of the cam and follower is called the *pressure angle*, as illustrated in Fig. 9-24. The only force that a cam can exert on a cam follower is

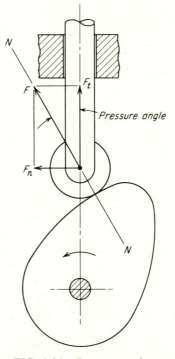

FIG. 9-24.   Pressure angle.

along their common normal. If this force is resolved into two components, one normal and one tangent to the follower path as shown in the figure, it is obvious that the normal force is undesirable and should be kept to a minimum. It is difficult to set an absolute maximum value for the pressure angle, since its importance is dependent upon many factors, such as cam speed, type of follower, coefficient of friction, resistance of the mechanism driven by the follower, etc., but the following rule is generally applied:

RULE

The pressure angle should be kept as small as possible and, in general, should not exceed 30°.

The pressure angle for a given displacement diagram can be reduced by one or more of the following methods:

1. Increase the diameter of the base circle.
2. Increase the angle of cam rotation for a given follower displacement.
3. Decrease the total displacement of the follower.
4. Change the amount of follower offset.
5. Change the type of follower motion, i.e., uniform-velocity, uniform-acceleration, etc.

9-10. Analysis of Cam-follower Motion

The methods presented in this chapter relate to the construction of displacement diagrams for cam-follower motions and the construction of the actual cam profiles. The analysis of the velocities and accelerations of cam followers has not been included because it would constitute a repetition of material covered in previous chapters. The velocities and accelerations of cam followers can be analyzed either by the vector methods explained in Chaps. 6 and 7 or by the graphical-differentiation methods presented in Chap. 8. If the cam-follower displacements are laid out according to one of the standard follower motions, such as simple harmonic motion, their velocity and acceleration characteristics are known in advance, thereby making a detailed velocity and acceleration analysis unnecessary in most cases.

Problems

Problems 9-1 to 9-10 consist in constructing displacement diagrams for the specifications given. Each diagram should be 6 in. long (½ in. = 30°), and its height should be equal to the full-size maximum follower displacement.

9-1. Uniform-velocity (straight-line) motion:
Rise 2 in. in 120°
Dwell 30°
Fall 1 in. in 90°
Dwell 30°
Fall 1 in. in 90° (modify with arc at each end)

9-2. Simple harmonic motion (SHM):
Rise 2 in. in 120°
Dwell 30°
Fall 2 in. in 180°
Dwell 30°

9-3. Uniformly accelerated (parabolic) motion (UARM):
Rise 2 in. in 180°
Dwell 30°
Fall 2 in. in 120°
Dwell 30°

9-4. Modified uniform-velocity (straight-line) motion. Each displacement should be modified by introducing uniformly accelerated portions at both ends. It is recommended that each displacement be divided into thirds: one-third uniform-acceleration, one-third uniform-velocity, and one-third uniform-deceleration.
Rise 2 in. in 180°
Dwell 45°
Fall 2 in. in 135°

9-5.   Cycloidal motion:
Rise 2 in. in 120°
Dwell 60°
Fall 2 in. in 180°

9-6.   Combination:
Rise 2 in. in 120° (modified uniform-velocity motion–uniform-acceleration method)
Fall 1 in. in 90° (simple harmonic motion)
Dwell 60°
Fall 1 in. in 60° (simple harmonic motion)
Dwell 30°

9-7.   Combination:
Rise 2 in. in 120° (cycloidal motion)
Dwell 60°
Fall 2 in. in 180° (simple harmonic motion)

9-8.   Combination:
Rise 2 in. in 120° (uniformly accelerated motion)
Dwell 30°
Fall 2 in. in 210° (simple harmonic motion)

9-9.   Combination:
Rise 2 in. in 90° (cycloidal motion)
Dwell 30°
Fall 1 in. in 90° (modified uniform-velocity motion—arc method)
Dwell 30°
Fall 1 in. in 120° (uniformly accelerated motion)

9-10. Combination:
Rise 2 in. in 180° (simple harmonic motion)
Dwell 60°
Fall 2 in. in 120° (uniformly accelerated motion)

Problems 9-11 to 9-15 consist in laying out the cam profile for the particular cam-and-follower arrangement shown. Use one of the cam-follower motions described in Probs. 9-1 to 9-10 as assigned by the instructor. These problems are designed to fit on an 8½- by 11-in. sheet with the short edge parallel to the T square. The center of the cam should be located at the center of the sheet.

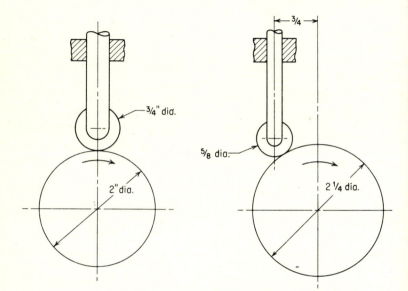

Fig. 9-25.   Prob. 9-11.   FIG. 9-26.   Prob. 9-12.

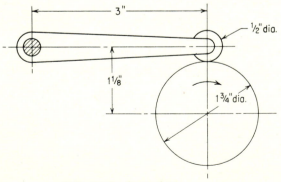

FIG. 9-27.   Prob. 9-13.

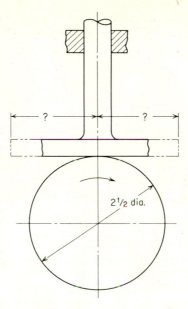

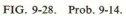

FIG. 9-28.   Prob. 9-14.

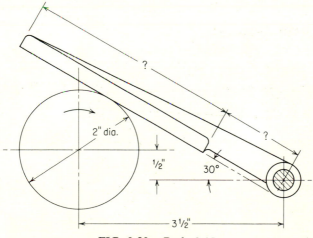

FIG. 9-29.   Prob. 9-15.

theory of gears

10-1. Introduction

Gears are toothed wheels used to transmit mechanical motion accurately, in the form of rotational movement, from one member to another. As can be seen in Fig. 10-1, gears can be designed in a wide variety of forms. Article 10-5 discusses the classification of these various forms.

The forerunner of gears was friction rollers, but rollers slip under load and do not maintain constant ratios or fixed angular relationships. Gears are to friction rollers what chains and sprockets are to belts and pulleys. Although friction rollers and belts are indispensable in some applications, they can only be used where some slippage can be tolerated.

10-2. Gear Requirements

Gears are very versatile machine elements. They range in size and use from tiny gears in instruments and watches to huge driving gears in punch presses. Regardless of their application, however, they must fulfill certain basic requirements if they are to function properly. They must (*a*) transmit motion (or power) smoothly, positively, and efficiently, (*b*) be capable of transmitting motion with constant angular velocity, (*c*) maintain fixed angular relationships between members,

FIG. 10-1.   Miscellaneous gears. (*Renold Crofts, Incorporated.*)

(*d*) be interchangeable with other gears having the same tooth size, and (*e*) be reasonably easy to manufacture.

10-3.   Action of Curved Surfaces in Direct Contact

The action between the gear teeth of two mating gears is the action of a pair of curved surfaces in direct contact and has the following characteristics: (*a*) the only motion that is transmitted from one member to the other is along their common normal, (*b*) the action that *generally* takes place is a combination of rolling and sliding, and (*c*) the angular velocities of the two members are inversely proportional to the segments into which their line of centers is divided by their common normal.

In the general case, the angular velocity ratio of the two members is continually changing. In Fig. 10-2, three positions of two curved members are shown. Figure 10-2*a* is the general case and depicts a

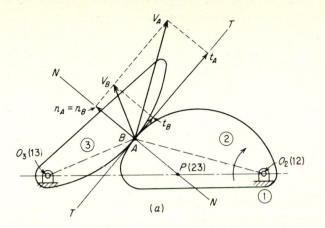

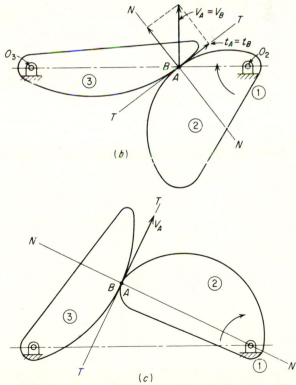

FIG. 10-2. Action of curved surfaces in direct contact. (a) Combination rolling and sliding. (b) Pure rolling: point of contact lies on line of centers. (c) Pure sliding: common normal passes through center of rotation of driver.

combination of rolling and sliding. *Unless the two members are separating or crushing, the normal components n_A and n_B of points A and B must be equal and lie along line NN, which is the common normal at the instant in question.* The actual velocities of A and B are not the same, nor are their directions the same. Their tangent components t_A and t_B must both lie in the direction of the common tangent TT, but they have different magnitudes and may even have opposite sense. The algebraic difference of these tangential components represents the amount of sliding that is taking place.

Figure 10-2b shows the point of contact lying on the line of centers of the two members. At this instant, the velocity vectors V_A and V_B coincide, and their tangential components are equal; therefore, no sliding is taking place. Thus, *when the point of contact of the two members lies on their line of centers, a condition of pure rolling exists.*

Figure 10-2c shows the members in a position in which the common normal NN passes through the center of the driving member. At this instant, the velocity of A is along the common tangent TT, and its normal component is zero; therefore, no motion is transmitted to B and $V_B = 0$. Thus, *when the common normal of the two members passes through the center of rotation of the driver, the motion is pure sliding.*

Referring again to Fig. 10-2a, point P is located at centro 23, which can be regarded as a point common to links 2 and 3 and has the same velocity in each; hence, the velocity of P can be expressed as

$$V_P = r\omega = (O_2P)\omega_2 = (O_3P)\omega_3$$

so

$$\frac{\omega_2}{\omega_3} = \frac{O_3P}{O_2P}$$

This shows that *the angular velocities of the two members are inversely proportional to the segments into which their line of centers is divided by their common normal.*

10-4. Action of Gear Teeth

As stated before, the study of gear-tooth action is fundamentally a study of the motion transmitted by a pair of members having curved surfaces (or at least surfaces convex to each other) in direct contact; in the general case there exists a combination of rolling and sliding, and the angular velocity ratio of the two members is continually changing.

One of the fundamental requirements of gears, however, is that they be capable of transmitting motion with constant angular velocity. Gear teeth, then, must have special shapes to accomplish this. It is possible to start with any shape of tooth for one gear and develop a conjugate curve for the mating gear tooth. For interchangeability, however, it is necessary that both gears have the same shape teeth. Kinematically, the involute curve is ideally suited for gear teeth, so most of the present-day gears are based on the involute. The only disadvantage of involute gear teeth is that, for gears having only a few teeth, the involute produces a weak, undercut tooth. For this reason, a modified involute is sometimes used.

An involute is generated by the end of a string being unwound from a cylinder or by a point on a line as the line rolls on the circumference of a circle without slipping. The characteristic of an involute that makes it ideally suited for a tooth form is that *any normal to an involute is tangent to the base circle from which the involute is generated.* Figure 10-3*a* shows an involute being generated by a straight line rolling on a circle. For any position of the line, the line is normal to the curve it is generating, and it is, of course, tangent to the circle. Figure 10-3*b* shows two involutes in contact. Notice that their common normal is tangent to *both* base circles; no matter where the contact point is, the common normal line will always be tangent to these base circles. If the centers of the two involutes are fixed, the common normal line will remain fixed and will always cross the line of centers at the same point. Therefore, the angular velocity ratio of the two involutes remains constant. This fulfills the second gear requirement in Art. 10-2, which is often referred to as the *law of gearing*.

The action of two gears having pure involute tooth forms is shown in Fig. 10-4, which shows two positions of the same mating teeth. The two teeth first make contact at *A*, and their point of contact travels along the *line of action* (common normal) *NN* until it reaches

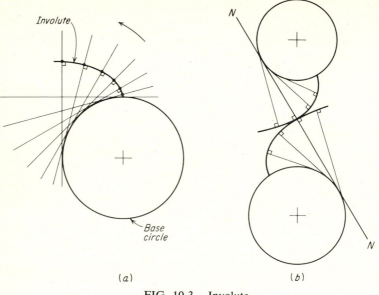

FIG. 10-3. Involute.

B, where contact is broken. The line AB, then, is the *path of contact*. Since the tooth form is involute, the common normal is always at the same angle and always passes through the same point on the line of centers O_2O_3. This point is the *pitch point P*. Thus, the gear ratio remains constant, as it must.

The circles 1 and 2 are the *base circles* for the involutes of the two gears. Notice that the line of action is the common normal with respect to the two mating gear teeth and that it is tangent to both base circles. The angle that this line of action NN makes with a line normal to the line of centers of the gears is called the *pressure angle*. Various pressure angles may be used, but certain angles have become standard. The smaller the pressure angle, the smaller the component tending to push the gears apart, but the weaker the tooth profile.

The circles 3 and 4 are the *pitch circles* for the two gears, and the *arc of action* is the arc along the pitch circle from the point on a

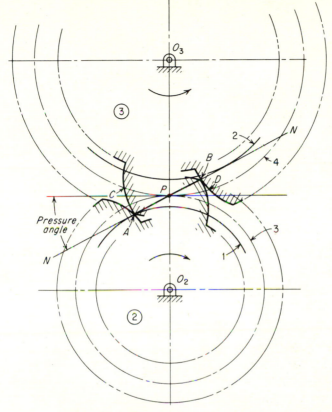

FIG. 10-4. Action of involute gear teeth.

tooth when it first makes contact to the same point when the tooth breaks contact. The arc of action for gear 3 in Fig. 10-4 is arc *CD*.

10-5. Classification of Gears

The term *gear* is often used in a general sense, meaning any toothed wheel that mates with another toothed wheel (as distinguished from a sprocket, which mates with a chain). In the strict sense, however, a *gear* is the larger of the two mating toothed wheels, and the smaller of the two is the *pinion*. In this text, the term gear or gears will frequently be used in the general sense. A *rack* can be thought of as a portion of a gear of infinite radius. It is conventional to classify gears as follows:

SPUR

This is the simplest form of gear for transmitting motion between two parallel shafts. A spur gear is shown in Fig. 10-5.

CLUSTER GEARS

Gear boxes or transmissions often utilize cluster gears consisting of several spur gears (or helical gears) cut on one shaft, as shown in Fig. 10-6.

FIG. 10-5. Spur gear. (*Browning Manufacturing Company.*)

FIG. 10-6. Cluster gear. (*Eaton Corporation.*)

HELICAL GEARS

These gears are similar to spur gears except that the teeth are helical rather than straight, as shown in Fig. 10-7. The curved teeth result in a smoother transmission of motion because each tooth picks up its load gradually. The disadvantage is that an undesirable side thrust is introduced. *Herringbone* gears, shown in Fig. 10-7c, have the smooth action of helical gears without their undesirable side thrust, but they are expensive to produce.

(a)

(b)

(c)

FIG. 10-7. Helical gears. (a) Single. (*FMC/Link-Belt.*) (b) Double. (*The Cincinnati Gear Company.*) (c) Herringbone. (*FMC/Link-Belt.*)

BEVEL GEARS

These gears are conical rather than cylindrical in form. Their main function is to transmit motion *around corners*. Bevel gears are produced in several forms, as illustrated in Fig. 10-8.

Straight bevel gears are the simplest form of bevel gears. They make an angle of 90° with each other, and their center lines intersect. They are comparable to spur gears in that their teeth are parallel to the elements of the cone (see Fig. 10-8*a*).

Spiral gears are an improvement over straight bevel gears in the same way that helical gears are an improvement over spur gears (see Fig. 10-8*b*).

Skew gears are bevel gears that make an angle *other* than 90° with each other and whose center lines do not necessarily intersect (see Fig. 10-8*c*).

Hypoid gears are bevel gears whose center lines are perpendicular but do not intersect. Their chief use is in connection with automotive rear axles, where the objective is to lower the drive shaft to permit car bodies to be built lower to the ground (see Fig. 10-8*d*).

The term *miter gears* is often used to describe a pair of bevel gears that are the same size and make an angle of 90° with each other.

WORM GEARS

A *worm* and *worm gear* set, commonly referred to simply as worm gears, is used primarily for obtaining a large speed reduction in a relatively small space (see Fig. 10-9). They also have a *self-locking* characteristic that is desirable in some applications. The worm can drive the worm gear in either direction, but, when it is not driving, the worm gear is locked; that is, the gear cannot drive the worm. This is especially true of the *single-lead* worm gear, which rotates only one tooth for each revolution of the worm. A *multiple-lead* worm gear advances two or more teeth for each revolution of the worm. Figure 10-9*b* is a quadruple lead worm and worm gear.

Figure 10-10 shows an internal or *annular gear* (and pinion), which has the advantages of reducing the required center distance and increasing the contact ratio for a given speed ratio as compared with a pair of spur gears.

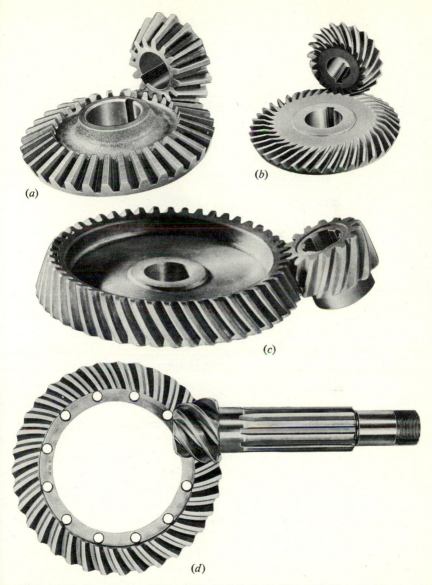

FIG. 10-8.　Bevel gears. (*a*) Straight. (*b*) Spiral. (*c*) Skew. (*d*) Hypoid. (*Eaton Corporation.*)

FIG. 10-9. Worm gears. (*a*) Single-lead. (*The Cincinnati Gear Company.*) (*b*) Multiple-lead. (*FMC/Link-Belt.*)

Figure 10-11 shows a set of *right-angle helical gears* and illustrates the versatility of helical gears. These gears can be made for transmitting motion between shafts that are parallel or that make any angle with each other. Right-angle helical gears are frequently used to drive an automotive distributor from the cam shaft.

FIG. 10-10. Annular gear and pinion. (*The Cincinnati Gear Company.*)

FIG. 10-11. Right-angle helical gears. (*Eaton Corporation.*)

10-6. Terms and Definitions

The terms and definitions presented in this section are limited to spur gears, although most of the terms are common to all types of gears. Once the theory of spur gears is understood, it is fairly easy to extend the methods to other types of gears.

The terminology for a single spur gear is shown in Fig. 10-12*a*, and most of the following terms are illustrated:

The *pitch diameter D*, sometimes referred to as the *effective diameter* of the gear, is the diameter of the *pitch circle*. The pitch

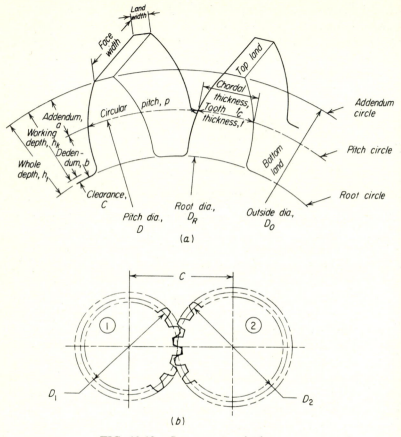

FIG. 10-12. Spur-gear terminology.

diameter forms the basis for all of the other gear relationships. The radius of the pitch circle is called the *pitch radius R.*

The *addendum a* is the top portion of the gear tooth and extends from the pitch circle to the *addendum circle.*

The *dedendum b* is the bottom portion of the gear tooth and extends from the pitch circle to the *root circle.*

The *clearance c* is the amount by which the dedendum is made larger than the addendum. It is also the amount by which the whole depth exceeds the working depth. Its purpose is to prevent the tops of mating gear teeth from *bottoming*.

The *whole depth* (or total depth or total height) h_t is the sum of the addendum and the dedendum.

The *working depth* (height) h_k is the depth to which the teeth of the mating gear penetrate. It is the whole depth less the clearance.

The *tooth thickness t* is the arc or circular thickness of the tooth measured along the pitch circle.

The *chordal thickness* t_c is the straight-line thickness of the tooth measured at the pitch circle. This is a convenient thickness for gauging purposes.

The *circular pitch p* is the distance measured along the pitch circle, from a point on one tooth to the corresponding point on the adjacent tooth.

The *outside diameter* D_o is the diameter of the addendum circle, and is the diameter of the gear *blank* prior to cutting the teeth.

The *center distance C*, as shown in Fig. 10-12*b*, is the center-to-center distance between two mating gears.

Diametral-pitch System. This is a system whereby the tooth size and all related gear computations are based on the *pitch diameter*.

The *diametral pitch P*, or simply *pitch*, of a gear is an expression of tooth size. It is *the number of teeth per inch of pitch diameter*. Thus, a gear with 2-in. pitch diameter and 20 teeth is a 10-*P* or 10-pitch gear, and a gear with a 2-in. pitch diameter and 40 teeth is a 20-pitch gear. Therefore, the larger the numerical value of *P*, the smaller the teeth. Any two gears having the same pitch will operate together, provided they are based on the same gear system (see Art. 10-8).

Figure 10-13 shows the sizes of the teeth for some common diametral pitches. Theoretically, it is possible to produce almost any size gear teeth, but in the interest of economy in tooling, the American Gear Manufacturers' Association (AGMA) lists certain preferred diametral pitches, as shown in Table 10-1. If values other than the recommended ones are used, preference is usually given to values which are *even integers*.

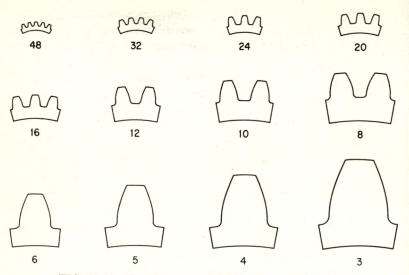

FIG. 10-13.   Standard gear-teeth sizes (diametral pitch).

Table 10-1.   RECOMMENDED
DIAMETRAL PITCHES

Coarse Pitch		Fine Pitch	
2	6	20	80
2.25	8	24	96
2.5	10	32	120
3	12	40	150
4	16	48	200
		64	

CIRCULAR-PITCH SYSTEM

This is a system whereby the tooth size and all related computations are based on the *circumference* of the pitch circle rather than on its diameter. This system was convenient in the days when gear teeth were laid out by hand, using dividers to step off the teeth around the gear. A simple or round number was chosen for the circular pitch. Because of the ratio of π between the circumference and diameter of a circle, if a simple number is used for circular pitch, the pitch diameters and center distance will be awkward numbers. With the advent of newer methods of producing gears, such as the use of automatic indexing devices that can easily divide any circle into any integral number of divisions, this system has become little used.

MODULE SYSTEM (BRITISH)

In this system, the gear tooth size is expressed as a *module*, which is the reciprocal of the diametral pitch, that is,

$$\text{Module} = \frac{1}{P} = \frac{D}{N}$$

Thus, a gear with a module of $\frac{1}{12}$ is the equivalent of a 12-pitch gear in the diametral-pitch system.

10-7. Basic Relationships

Since diametral pitch is the number of teeth per inch of pitch diameter, it can be expressed as

$$P = \frac{N}{D} \tag{10-1}$$

where P = diametral pitch
N = number of teeth
D = pitch diameter, in.

The addendum has been arbitrarily established as the reciprocal of the diametral pitch; therefore,

$$a = \frac{1}{P} \tag{10-2}$$

The clearance has been arbitrarily established as[1]

$$c = \frac{0.25}{P} \tag{10-3}$$

[1] Based on the standard coarse-pitch-gear system. The clearance for standard fine-pitch gears (20-pitch and smaller) is $0.200/P$ (see Table 10-2).

where c = clearance, in.

$\quad\quad P$ = diametral pitch

The dedendum is larger than the addendum by the amount of the clearance; therefore, it can be expressed as

$$b = a + c \tag{10-4}$$

where a = addendum, in.

$\quad\quad b$ = dedendum, in.

$\quad\quad c$ = clearance, in.

and substituting in Eqs. (10-2) and (10-3) gives

$$b = \frac{1.25}{P} \tag{10-5}$$

where P = diametral pitch.

From Fig. 10-12a it is obvious that the whole (total) depth of a gear tooth is equal to the sum of the addendum and the dedendum; so

$$h_t = a + b \tag{10-6}$$

where h_t = whole depth, in.

$\quad\quad a$ = addendum, in.

$\quad\quad b$ = dedendum, in.

and substituting in Eqs. (10-2) and (10-5) gives

$$h_t = \frac{2.25}{P} \tag{10-7}$$

where P = diametral pitch.

Again, from Fig. 10-12a it is obvious that the working depth of a gear tooth is twice the addendum, so that

$$h_k = \frac{2}{P} \tag{10-8}$$

where h_k = working depth, in.

$\quad\quad P$ = diametral pitch

The outside diameter of a gear is equal to the pitch diameter plus twice the addendum; therefore,

$$D_o = D + 2a \qquad (10\text{-}9)$$

where D_o = outside diameter, in.
D = pitch diameter, in.
a = addendum, in.

and substituting in Eqs. (10-1) and (10-2) gives

$$D_o = \frac{N + 2}{P} \qquad (10\text{-}10)$$

where N = number of teeth
P = diametral pitch
The outside radius of a gear is equal to one-half the outside diameter; so

$$R_o = \frac{D_o}{2} \qquad (10\text{-}11)$$

where R_o = outside radius, in.
D_o = outside diameter, in.
The pitch radius of a gear is equal to one-half the pitch diameter; so

$$R = \frac{D}{2} \qquad (10\text{-}12)$$

where R = pitch radius, in.
D = pitch diameter, in.

Since the diametral-pitch system is based on the pitch *diameter*, and the circular-pitch system is based on the pitch *circumference*, their relationship may be expressed as

$$pP = \pi \qquad (10\text{-}13)$$

where p = circular pitch, in.
P = diametral pitch

It is evident in Fig. 10-12*a* that if there were zero backlash,[1] the tooth thickness would equal one-half the circular pitch; therefore, the theoretical tooth thickness is expressed as

$$t = \frac{p}{2} \tag{10-14}$$

where t = tooth thickness, in.
p = circular pitch, in.
and substituting in Eq. (10-13) gives

$$t = \frac{\pi}{2P} \tag{10-15}$$

where P = diametral pitch.

It should be evident that the *actual* tooth thickness would be reduced by the amount of the backlash. Also, it should be noted that this is an *arc* tooth thickness.

From Fig. 10-12*b* it is evident that the center distance between two mating gears is[2]

$$C = \frac{D_1 + D_2}{2} \tag{10-16}$$

where C = center distance, in.
D_1 = pitch diameter of first gear, in.
D_2 = pitch diameter of second gear, in.
and substituting in Eq. (10-1) gives

$$C = \frac{N_1 + N_2}{2P} \tag{10-17}$$

where N_1 = number of teeth on first gear
N_2 = number of teeth on second gear
P = diametral pitch

---

[1] See Art. 10-10 for explanation of backlash.

[2] In the case of *annular* gears, it is the *difference* in the pitch diameters divided by 2.

CHORDAL TOOTH THICKNESS AND CHORDAL ADDENDUM

Gear teeth are frequently measured by vernier calipers (or fixed-caliper-type "go–no go" gauges), as shown in Fig. 10-14a. The addendum setting for the calipers has to be made slightly larger than the gear addendum, and the thickness setting has to be slightly smaller than the *circular* tooth thickness. To gauge gear teeth in this manner, then, it is necessary to calculate a *chordal* addendum and a *chordal* thickness.

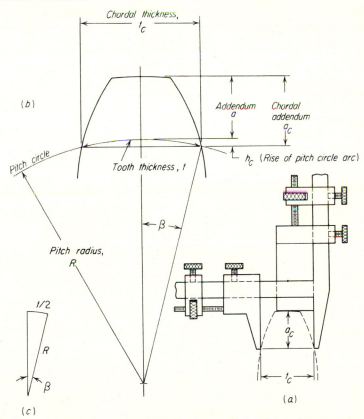

FIG. 10-14. Chordal addendum and chordal thickness.

In Fig. 10-14b, it is evident that the chordal addendum is

$$a_c = a + h_c \tag{10-18}$$

where a_c = chordal addendum, in.
   a = addendum, in.
   h_c = rise of the arc of the pitch circle, in.

The central angle subtended by an arc equal to half the tooth thickness (Fig. 10-14c) can be expressed in radians as

$$\beta = \frac{t/2}{R} = \frac{t}{2R} = \frac{t}{D} \tag{10-19}$$

where β (beta) = angle subtended by the half-tooth thickness, radians
   t = tooth thickness, in.
   R = pitch radius, in.

The above expression written in terms of degrees rather than radians (π radians = 180°) appears as

$$\beta = \frac{90t}{\pi R} \tag{10-20}$$

where β = angle subtended by the half-tooth thickness, deg
   t = tooth thickness, in.
   R = pitch radius, in.

It should be emphasized that Eq. (10-20) is valid for *any* tooth thickness. A much simpler expression can be used if the tooth thickness is exactly equal to half the circular pitch (zero backlash). In this case, the angle subtended by $t/2$ is 360°/4N or 90°/N.

The amount of rise of the arc of the pitch circle, as shown in Fig. 10-14b, is given by the expression

$$h_c = R(1 - \cos \beta) \tag{10-21}$$

where h_c = rise of pitch circle arc, in.
   R = pitch radius, in.
   β = angle subtended by half-tooth thickness, deg
and substituting Eq. (10-20) into Eq. (10-21) produces the following expression:

$$h_c = R\left(1 - \cos \frac{90t}{\pi R}\right) \tag{10-22}$$

Then, substituting Eq. (10-22) into Eq. (10-18) produces the expression

$$a_c = a + R\left(1 - \cos \frac{90t}{\pi R}\right) \qquad (10\text{-}23)$$

The *chordal thickness* is the length of the chord corresponding to the arc representing the tooth thickness, as shown in Fig. 10-14b, and may be expressed as

$$t_c = 2R \sin \beta \qquad (10\text{-}24)$$

where t_c = chordal thickness, in.
   R = pitch radius, in.
   β = angle subtended by half-tooth thickness, deg
and substituting Eq. (10-20) into Eq. (10-24) produces the following expression:

$$t_c = 2R \sin \frac{90t}{\pi R} \qquad (10\text{-}25)$$

Equations (10-23) and (10-25), then, are useful expressions for calculating the chordal addendum and chordal thickness, respectively.[1]

The following two examples will serve to illustrate the use of the above relationships:

EXAMPLE 10-1.   SPUR GEAR

A spur gear has 24 teeth and a diametral pitch of 4. Determine its (a) pitch diameter, (b) addendum, (c) clearance, (d) dedendum, (e) whole depth, (f) working depth, (g) outside diameter, (h) circular pitch, (i) tooth thickness, (j) chordal addendum, and (k) chordal thickness.

(a) $D = \dfrac{N}{P} = \dfrac{24}{4} = 6$ in.

(b) $a = \dfrac{1}{P} = \frac{1}{4}$ in.

---

[1] For a table of chordal addendum and chordal thickness values see Darle W. Dudley (ed.), *Gear Handbook*, pp. 24–33, McGraw-Hill Book Company, New York, 1962.

(c) $c = \dfrac{0.25}{P} = \dfrac{0.25}{4} = 0.063$ in.

(d) $b = a + c = \frac{1}{4} + 0.063 = 0.313$ in.

(e) $h_t = a + b = 0.250 + 0.313 = 0.563$ in.

(f) $h_k = 2a = (2)(\frac{1}{4}) = \frac{1}{2}$ in.

(g) $D_0 = D + 2a = 6 + (2)(\frac{1}{4}) = 6\frac{1}{2}$ in.

or

$$D_o = \frac{N + 2}{P} = \frac{26}{4} = 6\frac{1}{2} \text{ in.}$$

(h) $p = \dfrac{\pi}{P} = \dfrac{\pi}{4} = 0.785$ in.

(i) $t = \dfrac{p}{2} = \dfrac{0.785}{2} = 0.393$ in.

or

$$t = \frac{\pi}{2P} = \frac{\pi}{8} = 0.393 \text{ in.}$$

(j) $R = \dfrac{D}{2} = \dfrac{6}{2} = 3$ in.

then

$$a_c = a + R\left(1 - \cos \frac{90t}{\pi R}\right)$$

$$= \frac{1}{4} + 3\left[1 - \cos \frac{(90)(0.393)}{\pi(3)}\right]$$

$$= \frac{1}{4} + 3(1 - \cos 3.75°)$$

$$= 0.25 + 0.0065$$

$$= 0.2565 \text{ in.}$$

(k) $t_c = 2R \sin \dfrac{90t}{\pi R}$

$$= (2)(3) \sin 3.75°$$

$$= (6)(0.0654)$$

$$= 0.3924 \text{ in.}$$

EXAMPLE 10-2.   PAIR OF SPUR GEARS

If the gear in Example 10-1 is mated with a second gear which has 16 teeth, what is their center distance?

$$C = \frac{N_1 + N_2}{2P} = \frac{24 + 16}{(2)(4)} = 5 \text{ in.}$$

or

$$C = \frac{D_1 + D_2}{2} = \frac{6 + 4}{2} = 5 \text{ in.}$$

where

$$D_2 = N_2/P = {}^{16}\!/_4 = 4 \text{ in.}$$

10-8.   Standard Systems of Gear-tooth Proportions

Standard systems of gear-tooth proportions provide a means of obtaining interchangeability. For gears to be interchangeable, the following conditions must exist:

1.   Their diametral pitches must be the same.
2.   Their pressure angles must be the same.
3.   Their addendums and dedendums must be the same.
4.   Their tooth thicknesses must be the same (equal to half the circular pitch).

Because of a nearly infinite variety of possible tooth proportions, it is most desirable to establish a limited number of standard systems. The standard systems agreed upon by the American Gear Manu-facturers' Association (AGMA) and the American Standards Association (ASA) set forth the various relationships regarding tooth thickness, addendum, working depth, and pressure angle. These relationships are given in Table 10-2.

The two basic systems are the *coarse-pitch system* for gears with teeth larger than 20-pitch (smaller values of P) and the *fine-pitch system* for gears with teeth that are 20-pitch or smaller (larger values of P).

The coarse-pitch system provides for both 20° and 25° pressure angle while the fine-pitch system is for 20° pressure angles only. The applicable AGMA and ASA standards are given at the bottom of the table (items 13 and 14).

Table 10-2.   BASIC   TOOTH   PROPORTIONS   OF   SPUR   GEARS (AGMA AND ASA STANDARD SYSTEMS)

Item Number	Item	Coarse Pitch (Larger than $20P$) Full Depth		Fine Pitch ($20P$ and Finer) Full Depth
1	Pressure angle, ϕ	20°	25°	20°
2	Addendum, a	$\dfrac{1}{P}$	$\dfrac{1}{P}$	$\dfrac{1}{P}$
3	Dedendum, b	$\dfrac{1.25}{P}$	$\dfrac{1.25}{P}$	$\dfrac{1.200}{P} + 0.002$ in.
4	Working depth, h_k	$\dfrac{2}{P}$	$\dfrac{2}{P}$	$\dfrac{2}{P}$
5	Whole depth, h_t (min.)	$\dfrac{2.25}{P}$	$\dfrac{2.25}{P}$	$\dfrac{2.200}{P} + 0.002$ in.
6	Tooth thickness, t (circular)	$\dfrac{\Pi}{2P}$	$\dfrac{\Pi}{2P}$	$\dfrac{1.5708}{P}$
7	Fillet radius (of basic rack), r_f	$\dfrac{0.3}{P}$	$\dfrac{0.3}{P}$	Not std.
8	Clearance, c (min.)	$\dfrac{0.250}{P}$	$\dfrac{0.250}{P}$	$\dfrac{0.200}{P} + 0.002$ in.
9	Clearance, c (shaved or ground teeth)	$\dfrac{0.350}{P}$	$\dfrac{0.350}{P}$	$\dfrac{0.350}{P} + 0.002$ in.
	Min. number of teeth, $N_{\min}$			
10	Pinion	18	12	18
11	Pair	36	24	—
12	Min. width of top land	$\dfrac{0.25}{P}$	$\dfrac{0.25}{P}$	Not std.
	Ref. standards			
13	AGMA	201.02	201.02	207.04
14	ASA			$B6.19$

OBSOLETE SYSTEMS

Some standard systems are now obsolete for new designs but must be referred to occasionally when redesigning existing machinery. These obsolete systems are as follows:

Brown and Sharpe System. This system was developed by the Brown and Sharpe Company to replace the cycloidal tooth system. The principal features of this system were later embodied in the *composite system.*

AGMA $14\frac{1}{2}°$ *Full-depth Involute System.* This system was excellent for gears having at least 32 teeth but resulted in severely undercut teeth for gears having fewer teeth.

AGMA $14\frac{1}{2}°$ *Composite System.* This system was based on the Brown and Sharpe system and is interchangeable with it.

Fellows 20° *Stub-tooth System.* The Fellows Gear Shaper Company developed this system in 1898. The purposes of the system were to achieve a stronger tooth profile and to permit fewer teeth on the pinion. The system reduced tooth interference by using the higher pressure angle and shorter teeth.

Cycloidal Tooth System. The cycloidal system was originated to avoid undercutting in pinions having small numbers of teeth. The cycloidal tooth profile is no longer used for any types of gears except clocks and timers.

Table 10-3 gives the basic tooth proportions for the more common obsolete gear systems. The AGMA standard 201.02A covers the two obsolete $14\frac{1}{2}°$ systems for reference purposes.

10-9. Nonstandard Gears

Standard gear proportions should be forsaken only in those situations where design limitations make the use of standard gears impractical or unsatisfactory. For economy in tooling and in stocking replacement gears, the use of nonstandard gears should be discouraged. Advantages can be gained, however, in certain cases by varying such things as pressure angle, tooth depth, addendum, or center distance. The main reasons for making these variations are to eliminate undercutting, to prevent interference, or to increase the contact ratio.

Table 10-3. BASIC TOOTH PROPORTIONS FOR OBSOLETE GEAR SYSTEMS

Item	AGMA 14½° Composite System and Brown and Sharpe System	Fellows 20° Stub System	AGMA Full-depth System
Pressure angle, ϕ	$14\frac{1}{2}°$	$20°$	$14\frac{1}{2}°$
Addendum, a	$\dfrac{1}{P}$	$\dfrac{0.8}{P}$	$\dfrac{1}{P}$
Dedendum, b	$\dfrac{1.157}{P}$	$\dfrac{1}{P}$	$\dfrac{1.157}{P}$
Working depth, h_k	$\dfrac{2}{P}$	$\dfrac{1.6}{P}$	$\dfrac{2}{P}$
Whole depth, h_t	$\dfrac{2.157}{P}$	$\dfrac{1.8}{P}$	$\dfrac{2.157}{P}$
Tooth thickness (circular), t	$\dfrac{\Pi}{2P}$	$\dfrac{\Pi}{2P}$	$\dfrac{\Pi}{2P}$
Fillet radius (of basic rack), r_f	$\dfrac{0.157}{P}$	Not std.	(1.33) (clearance)
Clearance,* c (min.)	$\dfrac{0.157}{P}$	$\dfrac{0.2}{P}$	$\dfrac{0.157}{P}$
Min. number of teeth, N_{min}			
Pinion	32	14	34
Pair	64	—	64
Ref. AGMA Standard	201.02A		201.02A

* Clearance must be adjusted for *ground* or *shaved* gears.

LONG- AND SHORT-ADDENDUM SYSTEM[1]

The AGMA has included in its standard gear systems provisions for modifying the addendum of gears to prevent undercutting where the number of teeth falls below the minimum, as shown in Table

[1] For a more complete discussion of this system, see Joseph E. Shigley, *Kinematic Analysis of Mechanisms*, 2d ed., pp. 276–277, McGraw-Hill Book Company, New York, 1969.

10-2. The improved action is obtained by "backing out" the cutter from the *pinion* blank, which results in a tooth profile farther away from the base circle. If the cutter is advanced a corresponding amount into the *gear* blank, the center distance, the pitch circles, and the pressure angle remain unchanged. The resulting gears, of course, are no longer interchangeable with standard gears.

10-10. Practical Aspects of Gear Design

In practical gear design there is always a pressure on the designer to make the design as compact as possible to save both space and weight. Any drastic reductions along these lines, however, are usually at the expense of performance; strength and life are decreased while noise and vibration are increased.

Four practical factors which must be given adequate attention if satisfactory performance is to be enjoyed are:

1. Backlash
2. Interference
3. Undercutting
4. Contact ratio

BACKLASH

This is the amount by which the width of a tooth space (measured along the pitch circle) exceeds the tooth thickness. It is the amount that a gear can turn without its mating gear turning. In most power-gearing situations some backlash is necessary to provide for lubrication, for production variations, and for thermal expansion. Each designer must decide how much backlash can be tolerated in a particular situation. Gears that frequently or suddenly reverse directions must have carefully controlled backlash. Backlash is difficult to control, however, because it is a function of pitch diameter, tooth thickness, and center distance. The effect of varying the center distance alone, for example, is given by the approximate expression

$$\Delta B \approx 2 \tan \phi (\Delta C) \tag{10-26}$$

where ΔB = change in backlash, in.

ΔC = change in center distance, in.

ϕ = pressure angle, deg (for $\phi = 20°$, $2 \tan \phi = 0.728$)

Table 10-4 gives recommended backlash values for various combinations of diametral pitch and center distance. The values in this table are relatively conservative and should result in satisfactory performance in most power applications. Stock gears (off the shelf)

Table 10-4. RECOMMENDED BACKLASH FOR POWER GEARING

Diametral Pitch	Center Distance, in.						
	0–5	5–10	10–20	20–30	30–50	50–80	80–120
½	—	—	—	—	0.045	0.060	0.080
1	—	—	—	0.035	0.040	0.050	0.060
2	—	—	0.025	0.030	0.037	0.045	0.055
3	—	0.018	0.022	0.027	0.035	0.042	—
4	—	0.016	0.020	0.025	0.030	0.040	—
6	0.008	0.010	0.015	0.020	0.025	—	—
8	0.006	0.008	0.012	0.017	—	—	—
10	0.005	0.007	0.010	—	—	—	—
12	0.004	0.006	—	—	—	—	—
16	0.004	0.005	—	—	—	—	—
20	0.004	—	—	—	—	—	—
32	0.003	—	—	—	—	—	—
64	0.002	—	—	—	—	—	—

usually must provide for considerably more backlash to permit maximum flexibility of application. Typical backlash values for these gears varies between $0.3/P$ and $0.5/P$. For example, a 10-pitch stock gear would provide for a backlash of 0.030 in., whereas Table 10-4 recommends 0.010 in. for 10-pitch gears mounted on, say, 12-in. centers.

INTERFERENCE

This phenomenon occurs when attempts are made to make gear assemblies too compact by shortening center distances or by using gears with too few teeth. Fortunately, it is easy to check gears for interference by the graphical method shown in Fig. 10-15. Point *A* in the figure represents the first contact of a particular pair of teeth, assuming that gear 2 is rotating clockwise, and point *B* represents

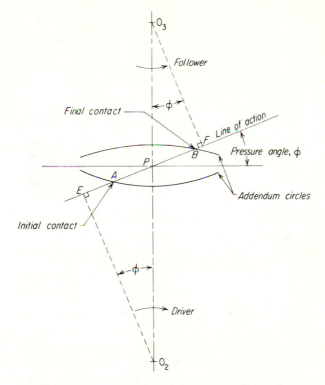

FIG. 10-15.   Graphical check for gear-tooth interference.

the last point of contact of the same two teeth. Line *AB* is the *path of contact* along the *line of action*. Points *E* and *F* are the tangent points of the line of action with the two base circles (see Art. 10-4), and they are referred to as the *interference points*. An interference exists if either point *A* or *B* lies outside of points *E* and *F*, respectively.

The result of interference is that the tips of the teeth on one of the gears scrapes against the flanks of the mating-gear teeth, causing a jerky motion, a breakdown of the lubrication film, and, ultimately, failure of the teeth surfaces.

Figure 10-16*a* depicts a situation where interference exists, for point *B* lies outside of point *F*. To determine the amount of the interference, draw a new *limiting* addendum circle for gear 2 through the interference point *F*. The radial difference between this limiting addendum circle and the original addendum circle is the amount of

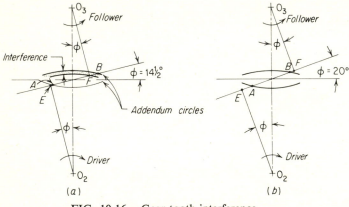

FIG. 10-16. Gear-tooth interference.

interference. Figure 10-16*b* shows that increasing the pressure angle to 20° eliminates the interference. This serves to emphasize one of the major advantages of the 20° pressure angle over the now-obsolete 14½° pressure angle. Interference, of course, can occur with *any* pressure angle if an attempt is made to use gears with too few teeth. Should interference be a problem, it can be eliminated by (*a*) increasing the pressure angle, (*b*) increasing the number of teeth on the gears by either using larger gears or by using smaller teeth, or (*c*) utilizing the long- and short-addendum system (see Art. 10-9). It should be pointed out that if the recommended pinion-teeth minima given in Table 10-5 are observed, interference should be no problem.

Table 10-5. MINIMUM NUMBER OF PINION TEETH VS.
PRESSURE ANGLE AND HELIX ANGLE TO OBTAIN NO
UNDERCUT

| | Min. No. of Teeth to Avoid Undercut | | | |
| | Normal Pressure Angle, ϕ_n | | | |
Helix Angle, deg	$14\frac{1}{2}°$	$20°$	$22\frac{1}{2}°$	$25°$
0 (Spur gear)	32	17	14	12
5	32	17	14	12
10	31	17	14	12
15	29	16	13	11
20	27	15	12	10
23	25	14	11	10
25	24	13	11	9
30	21	12	10	8
35	18	10	8	7
40	15	8	7	6
45	12	7	5	5

UNDERCUTTING

An undercut gear tooth is one in which a portion of the active
profile has been removed by the generating process. It results in an
unsatisfactory gear for two reasons. First, the tooth is weakened by
the undercutting since the tooth thickness often becomes smaller at
the root circle than at the pitch circle. Secondly, since the undercutting destroys part of the active surface, contact with the mating
gear is prevented over the affected portion, which results in increased
backlash, noise, vibration, and decreased life. As in the case of interference, undercutting is the result of attempting to make the gear
train too compact. In fact, one way to eliminate interference is by
deliberately undercutting the gear teeth, but this is not usually a
satisfactory solution.

Tables 10-3 and 10-5 contain information regarding the
minimum number of teeth permissible to avoid undercutting.

CONTACT RATIO

This is the average number of teeth in contact at any instant. It is obvious that the limiting value of the contact ratio is 1 and the practical minimum value is considered to be about 1.4 or 1.5. Higher contact ratios result in smoother action, increase the power-transmitting ability of the gears (more teeth share the load), and prolong the gear life.

The contact ratio can be expressed in two ways: either as the ratio of the arc of action (arc *CD* in Fig. 10-4) and the *circular pitch* (see Fig. 10-12*a*), or as the ratio of the length of the *path of contact* (*AB* in Fig. 10-15) to the *base pitch*. This last ratio is a very convenient one, and may be written in equation form as

$$m_p = \frac{Z}{p_b} \tag{10-27}$$

where m_p = contact ratio
Z = length of path of contact, in.
p_b = base pitch, in.

The *base pitch* p_b is the constant distance that exists between the corresponding profiles of two adjacent involute teeth as measured along any common normal (the line of action, for example) *or* as measured along their base circle. Figure 10-17 shows that two adjacent involutes generated from the same base circle have a constant distance measuring along any common normal such as N_1, N_2, or N_3 (that is, $AC = DE = FG$). Also, since an involute is generated by an "unwrapping" of the circumference of the generating base circle, arc length *AB* is equal to the straight-line distance *AC*.

Figure 10-18 shows the relationship that exists between the base pitch p_b and the circular pitch p. In the special case represented by a *rack*, where the pitch circle and the circular pitch become straight-line, this relationship is more clearly evident. The base pitch is shown to be the tooth-to-tooth distance measured along the line of action and the circular pitch is the tooth-to-tooth distance measured along the "pitch circle" of the rack. The base pitch and the circular pitch, then, are related as follows:

$$p_b = p \cos \phi \tag{10-28}$$

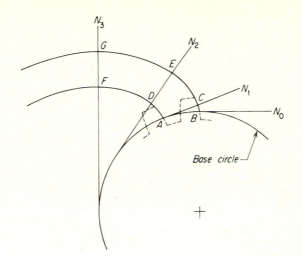

FIG. 10-17. Constant distance between adjacent involutes.

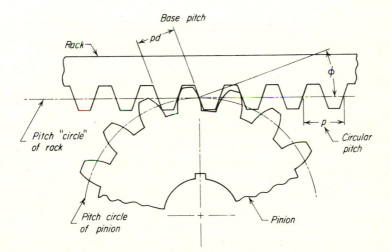

FIG. 10-18. Relationship between base pitch and circular pitch.

where p_b = base pitch, in.

p = circular pitch, in.

ϕ = pressure angle, deg

Also, this relationship can be written in terms of *diametral* pitch utilizing Eq. (10-13), or

$$p_b = \frac{\pi \cos \phi}{P} \qquad (10\text{-}29)$$

where P = diametral pitch.

Substituting this expression for p_b into Eq. (10-27) provides the convenient equation

$$m_p = \frac{Zp}{\pi \cos \phi} \qquad (10\text{-}30)$$

where m_p = contact ratio

Z = length of path of contact, in.

P = diametral pitch

ϕ = pressure angle, deg

The length of the *path of contact*, Z, can best be obtained graphically as shown in Fig. 10-19. All that is required to make the layout are the center distance, the outside radii of the two gears, the pitch radius of one of the gears (to locate pitch point P), and the pressure angle. In the figure, points A and B are the initial and final contact points, respectively, and the distance AB is the length of the path of contact. Notice that if the senses of the gears were reversed, the figure would be reversed but the length of the path of contact would be the same. Therefore, if the senses are unknown, or changing, they may be assumed.

EXAMPLE 10-3

Determine the contact ratio for a pair of mating spur gears whose diametral pitch is 4, whose pressure angle is 20°, and whose pitch diameters are 4 and 6 in. (same as gears used in Examples 10-1 and 10-2).

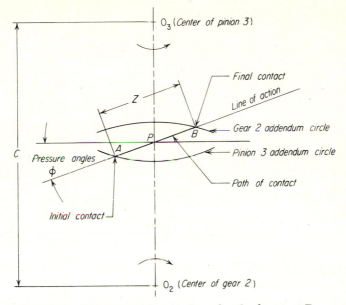

FIG. 10-19. Graphical determination of path of contact Z.

SOLUTION

1. Determine their center distance.

$$C = \frac{D_2 + D_3}{2} = \frac{4 + 6}{2} = 5 \text{ in.}$$

2. Determine the addendum.

$$a = \frac{1}{P} = \frac{1}{4} \text{ in.}$$

3. Make a graphical layout to determine the length of their path of contact, as shown in Fig. 10-20.

    a. Lay out points O_2 and O_3, 5 in. apart.

    b. Locate the pitch point P on the line of centers O_2–O_3 so that it is 2 in. from O_2.

    c. Lay out the common tangent through P and perpendicular to the line of centers O_2–O_3.

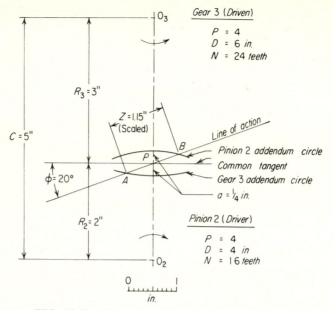

FIG. 10-20. Example of determining path of contact.

d. Lay out the line of action through the pitch point P and at $20°$ to the common tangent. For layout purposes, it is necessary to assume a sense for gear 2 (clockwise in this case).

e. Lay out the addendum circles for the two gears where the outside radii are

$$R_{o_2} = \frac{D_2}{2} + a = \frac{4}{2} + \frac{1}{4} = 2\frac{1}{4} \text{ in.}$$

$$R_{o_3} = \frac{D_3}{2} + a = \frac{6}{2} + \frac{1}{4} = 3\frac{1}{4} \text{ in.}$$

f. The points where the two addendum circles (outside radii) intersect the line of action determine the length of the path of contact, Z, which is scaled and found to be 1.15 in.

g. Calculate the contact ratio using Eq. (10-30).

$$m_p = \frac{ZP}{\pi \cos \phi} = \frac{(1.15)(4)}{\pi(0.9397)} = 1.56$$

Since it is desirable for the contact ratio to be at least 1.4, this contact ratio is satisfactory.

10-11. Gear Specifications

The specifications needed to completely describe a gear for production purposes are usually presented in the form of a detail drawing of the gear, which includes the following three general types of information:

1. Gear-blank dimensions, including outside diameter, face width, hub diameter, hub length, hole diameter, and details of fastening devices such as setscrews or keyways. This information is generally given on the actual views of the gear, in the form of dimensions.

2. Gear-tooth data which are usually given in tabular form and include number of teeth, diametral pitch, pressure angle, pitch diameter, circular thickness, whole depth, working depth, and occasionally chordal addendum and chordal thickness. In addition, there may be included reference information pertaining to the mating gear, or to the gear assembly, such as number of teeth on mating gear, part number of mating gear, center distance, and backlash.

3. Finally, there is a host of detailed information, usually given in the form of general notes or title-block "fill ins" pertaining to such things as material, heat treatment, protective coating, tolerances, surface quality (surface roughness), and geometric requirements (roundness, parallelism, runout, etc.).

The third group of specifications is the most difficult one to generalize about since it is highly dependent upon the service requirements imposed by the general type of equipment on which the gears are to be used, and by company or industry standards. In the past, gears have been classified by such terms as "commercial," "precision," "high accuracy," and "aircraft" quality. The latest AGMA standard (AGMA 390.01 May 1961) has classified all gear work into three broad ranges, which are listed on the next page.

COMMERCIAL-QUALITY RANGE

Gears are cut or formed by the less precise methods. They are either made oversize for the particular application or they are made of materials which permit a certain amount of "wearing-in" after installation.

PRECISION-QUALITY RANGE

Gears are cut or formed on precision equipment and are usually given a subsequent finishing operation such as shaving, grinding, or lapping.

ULTRAPRECISION-QUALITY RANGE

These gears are cut and finished by the most accurate processes the gear "art" can provide. They are usually assembled with little or no backlash.

The AGMA has assigned quality numbers with associated recommended tolerances. Quality classes 3 to 7 cover the general range of commercial quality; classes 8 to 12 comprise the precision range; and classes 13 to 15 are considered to be the ultraprecision range.

There are, of course, many tolerances associated with gears, and the proper application of tolerances is most complex. Any standards, such as those published by AGMA, are to be used only as guides; each designer must rely on his own judgment and experience to a large degree. Moreover, the determination of tolerances is not within the province of kinematics. With these thoughts in mind, the following *typical* tolerances are offered only as a sample:

Table 10-6. CENTER-HOLE TOLERANCES

Item	Fine-pitch Series		Coarse-pitch Series	
	Commercial	Precision	Commercial	Precision
Bore	+0.0010	+0.0002	+0.0015	+0.0007
Shaft	−0.0005	−0.0002	−0.0010	−0.0003
Min. clearance	0.0003	0.0000	0.0010	0.0002

Table 10-6 gives some typical tolerances for bore and shaft diameters by quality range. In applying these values, it is conventional to use the *basic-hole* philosophy of tolerance whereby the shaft is made undersize to provide the minimum clearance (allowance) and the tolerances are applied unilaterally in the direction of looser fit. For example, a fine-pitch, commercial-quality gear with a nominal bore size of $\frac{1}{2}$ in. would show a dimension of 0.5000 + 0.0010 and the shaft for this gear would be dimensioned 0.9997 − 0.0005.

Table 10-7 gives some typical center-distance tolerances by quality range.

Table 10-7. CENTER-DISTANCE TOLERANCES*

Quality	< 1 in.	1–6 in.	6–12 in.	12–24 in.
Commercial	0.002	0.003	0.005	0.010
Precision	0.0005	0.001	0.002	0.002
Ultraprecision	0.0001	0.0002	0.0002	

* All values are ±.

Table 10-8 gives some typical surface-quality (surface roughness) value ranges for the three quality levels.

Figure 10-21 shows a typical detail drawing of a spur gear. Notice that the gear teeth are not shown. The addendum and root circles are shown as phantom (double dash) lines and the pitch circle is shown as a center (single dash) line. Occasionally, one or two teeth

Table 10-8. TOOTH-SURFACE QUALITY

Quality	Roughness Height (μ in., rms)
Commercial	63–250
Precision	8–32
Ultraprecision	4–16

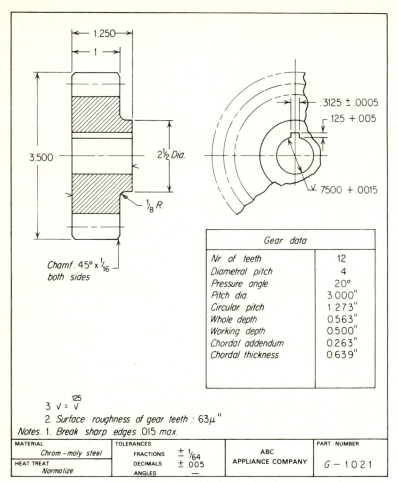

FIG. 10-21. Typical gear drawing.

are drawn to show their approximate size and shape, but usually not. In the cases where it is desirable to actually show the teeth (parts manuals and occasionally on assembly drawings), an approximate method for drawing gear teeth is used. These methods are described in most engineering-graphics texts.

Problems

10-1. What is the pitch diameter of a 10-pitch gear having 31 teeth?

10-2. An 8-pitch gear has a 6-in. pitch diameter. Find its (*a*) number of teeth (*b*) addendum and dedendum (*c*) outside diameter (*d*) circular pitch (*e*) tooth thickness (*f*) chordal addendum and (*g*) chordal thickness.

10-3. A gear is measured and found to have an outside diameter of 3.25 in., and it has 24 teeth. Find (*a*) the diametral pitch (*b*) the pitch diameter (*c*) the addendum (*d*) the dedendum and (*e*) the clearance.

10-4. Two 10-pitch gears are to operate together. If one of the gears has 50 teeth and the other has 25 teeth, what must be their center distance when they are mounted?

10-5. Some spur gears are to be cut from round stock. Their pitch diameter is to be $3\frac{1}{2}$ in., and their diametral pitch is 12. What diameter bar stock should be ordered if $\frac{1}{8}$ in. (0.125) on the diameter must be provided for cleaning up and truing? Stock is available in $\frac{1}{4}$-in. increments.

10-6. If a 32-pitch gear has an outside diameter of 3.0625, what is its pitch diameter and whole tooth depth?

10-7. A 6-pitch gear has 32 teeth. Find its (*a*) pitch diameter (*b*) outside diameter (*c*) addendum (*d*) clearance (*e*) dedendum (*f*) tooth thickness and (*g*) circular pitch.

10-8. Two 20° pressure-angle 10-pitch gears are mounted 8 in. on centers. If their backlash is established per Table 10-4, and if the tolerance on their center distance is ± 0.006, what is the range of the backlash values to be expected?

10-9. Two mating 4-pitch gears have 12 and 42 teeth, respectively. Find their (*a*) pitch diameters (*b*) mounting or center distance (*c*) outside diameters (*d*) whole depth and (*e*) tooth thickness.

10-10. Two mating gears have 56 and 128 teeth, respectively, and their center distance is 5.750 in. Find their (*a*) diametral pitch (*b*) pitch diameters (*c*) total tooth depth and (*d*) tooth thickness.

10-11. The outside diameter of an 8-pitch gear is 6.250 in. Find (*a*) the pitch diameter and (*b*) the number of teeth.

10-12.   A 20-tooth pinion drives a 100-tooth internal or annular gear. If the gears are 5-pitch, what must be their center distance?

10-13.   A pair of 6-pitch, 20° pressure-angle spur gears have 12 and 21 teeth, respectively. Check the pair for tooth interference.

10-14.   A pair of 8-pitch, 20° pressure-angle spur gears have 22 and 30 teeth, respectively. Determine their contact ratio.

10-15.   Determine the caliper settings, depth and width, to check tooth thickness of a 10-pitch gear with 24 teeth.

10-16.   Make a detail drawing of a 10-pitch, 20° pressure-angle gear, with a pitch diameter of 4 in., a face width of $\frac{3}{4}$ in., a bore of $1\frac{1}{4}$ in., a $\frac{5}{16}$-in. keyway, and a hub length of $1\frac{1}{2}$ in. centered with the face.

trains: belt, chain, and gear

11-1. Trains

With respect to mechanisms or machines, a *train* may consist of any arrangement or combination of links, belts, chains, gears, friction surfaces, pulleys, or cables. The most important feature of the train, from a kinematic standpoint, is the ratio of the *output* angular velocity to the *input* angular velocity, which is known as the *velocity ratio* (or *train value* in the case of gear trains). Theoretically, a train may consist of only two elements, such as a pair of gears, but generally a train is understood to consist of three or more elements. Although the most common usage of the word train is in connection with gears, a brief explanation of two other important elements, belts and chains, is included here because of their close relationship to gears.

11-2. Belts

Belts are used in connection with pulleys or sheaves to transmit rotary motion smoothly, quietly, and inexpensively. Their chief disadvantage is that they do not provide a positive drive: except in the case of toothed belts, there always exists some slippage or creep. This slipping ability is often an advantage in that it enables the belt to function as a shock absorber or as an overload protector. In fact, a belt can be alternately tightened and loosened to function as a friction clutch.

Technically, *slippage* consists of an actual sliding of the belt with respect to the sheave and is a function of the load, the belt tension, and the coefficient of friction existing between the belt and the sheave. *Creep*, on the other hand, occurs because the belt is under greater tension when it first makes contact with the driving sheave than it is when it leaves the sheave. In other words, the sheave delivers a shorter length of belt than it receives. Creep is largely dependent on the elasticity of the belt and on the load.

Belts are available in a variety of sizes and cross sections, including flat, round, and V sections, but V belts are by far the most common type. To facilitate interchangeability, V-belt manufacturers have developed industry standards, as shown in Fig. 11-1. Although the exact dimensions, shapes, and method of manufacture vary from one company to another, all will operate interchangeably in standard sheaves. The three broad categories are *industrial, agricultural*, and *automotive*. The industrial group consists of (*a*) the *conventional* sizes, which have long been used for the entire range of applications; (*b*) the newer *narrow* series, which cover the same range of applications as the conventional but are more compact; (*c*) the *light-duty* or fractional-horsepower (FHP) series, which is widely used in the fan and appliance fields; and (*d*) the *double V* series, which are available in the same sizes as the conventional belts. The agricultural and automotive series are especially designed to take care of the requirements peculiar to these areas.

Figure 11-2 shows some typical V-belt installations, including ganged belts on multigrooved sheaves. By using more than one belt, the power capacity of the installation can be improved by several orders of magnitude with very little increase in space requirements.

Figure 11-3 shows the typical construction of several standard V belts. Figure 11-3*a* shows a 3V belt with a cog-type construction which increases its ability to operate around small sheaves and with small center distances. Figure 11-3*b* shows the middle-size 5V belt and Fig. 11-3*c* shows the large, heavy-duty 8V belt where strength and stretch resistance are more important than flexibility. Figure 11-3*d*, of course, is the double-V belt for applications where the back of the belt must also operate on a sheave. Notice that, in all cases, the belts

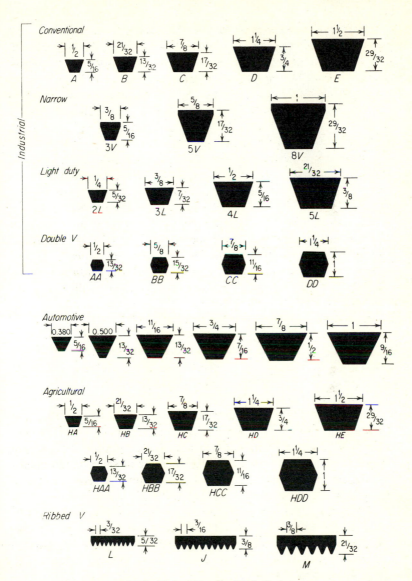

FIG. 11-1. Standard V-belt sizes.

FIG. 11-2.   V-belt installations. (*a*) Single. (*b*) Triple with adjustable-diameter sheave. (*Browning Manufacturing Company*.) (*c*) Five-groove sheaves. (*Gates Rubber Company*.)

consist of a flexible rubber core containing longitudinal cords for strength and encased in an impregnated fabric for optimum friction and wear resistance.

Figure 11-4 shows a special wide belt used for variable-speed belt drives. The cog-belt design increases the flexibility, and the crowned cross section increases lateral strength, both of which are extremely important in variable-speed applications. Figure 11-5 shows this type of belt used on two variable-speed drive units. In Fig. 11-5*a*, the spring-loaded variable-size sheave changes size as the position of the motor is changed. In Fig. 11-5*b*, the center distance remains constant and both sheaves change diameter, providing a much greater range of velocity ratios.

Figure 11-6 shows a link-type V belt which is particularly useful in situations where a regular belt could not be installed due to shaft

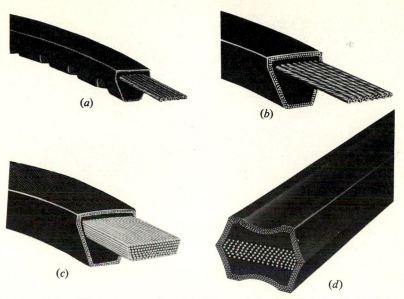

(*a*)

(*b*)

(*c*)

(*d*)

FIG. 11-3. V-belt construction. (*a*) 3V cog belt. (*b*) 5V belt. (*c*) 8V belt. (*d*) Double-V belt. (*The Goodyear Tire & Rubber Company.*)

FIG. 11-4. Special V belt for variable-speed applications. (*The Goodyear Tire & Rubber Company.*)

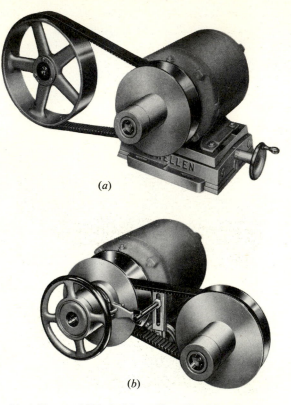

(a)

(b)

FIG. 11-5.   Variable-speed V-belt drives. (*Lewellen Manufacturing Company*.)

FIG. 11-6.   Link-type V belt. (*Gates Rubber Company*.)

extensions or other obstructions. Also, it can be made to any length and can be quickly repaired or changed in length.

Figure 11-7 shows two types of belts used where the power requirements are high and yet flexibility is important. The ribbed belt shown in Fig. 11-7a is available with up to 20 ribs for transmitting up to 1,700 horsepower. The integrally ganged V belts in Fig. 11-7b are available with two to five units in *conventional* sizes B, C, and D and in all three narrow sizes. Figure 11-8 shows both of these belt types installed on sheaves.

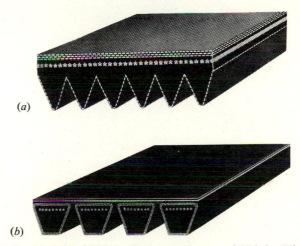

(a)

(b)

FIG. 11-7.   (a) Ribbed V belt. (b) Integrally ganged V belt. (*The Goodyear Tire & Rubber Company.*)

Figure 11-9 shows two types of positive-drive (or toothed) belts. These belts are becoming quite popular because they combine many of the advantages of belts (quiet, resilient, and lubrication-free) with at least one of the major advantages of a chain drive—the ability to maintain positive angular relationships. The dual belt shown in Fig. 11-9b has the ability to drive a sheave from either side. Figure 11-10 shows a positive-drive belt mounted on the special toothed sheaves. One relatively new use for this type of belt is in driving overhead cam shafts in automotive engines, where quietness and long life are both important.

In designing a belt drive, it is necessary to know the desired velocity ratio and the space limitations. It is also necessary to know the torque requirements in order to determine the size and number of belts, but this is in the realm of machine design and is not considered in this text.

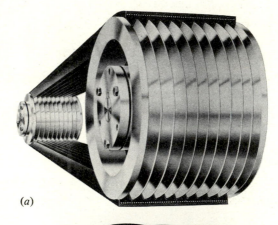

(*a*)

(*b*)

FIG. 11-8. (*a*) Ribbed V-belt installation. (*Browning Manufacturing Company*.) (*b*) Integrally ganged V-belt installation. (*Gates Rubber Company*.)

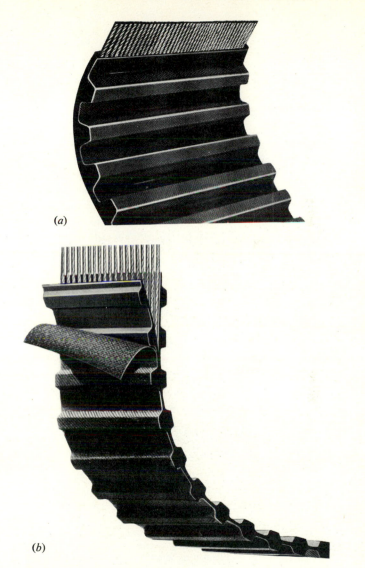

(a)

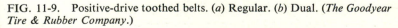

(b)

FIG. 11-9.   Positive-drive toothed belts. (*a*) Regular. (*b*) Dual. (*The Goodyear Tire & Rubber Company.*)

FIG. 11-10. Positive-drive belt installation. (*Browning Manufacturing Company.*)

The velocity ratio of two sheaves connected by belt is expressed as [1]

$$VR = \frac{\omega_F}{\omega_D} = \frac{D_D}{D_F} \tag{11-1}$$

where ω_F, ω_D = angular velocities of follower and driver, expressed in either radians or revolutions per unit of time [2]

D_D, D_F = diameters (pitch diameters), in.

This expression neglects slippage and creep, which can amount to as much as 5 percent of the theoretical speed, under heavy-load conditions.

Once the pulley sizes have been selected, it is necessary to compute the length of belt required for the desired center distance. The closest standard length is then selected from a catalogue or handbook, and the center distance is altered, if necessary, to agree with the available belt length. In addition, it is necessary to provide an adjustment in the center distance to permit the pulleys to be moved closer together for installing the belt and farther apart to compensate for stretch and wear.

[1] The inverse relationship existing between the angular velocities of the diameters of the drivers and followers is explained later, in Art. 11-5.

[2] In the earlier chapters, *n* is used in place of ω when the angular velocity is expressed as rpm. In this chapter ω is used for all angular velocities to avoid confusion, since N is used for number of gear teeth.

In most cases, the quickest way to determine the belt length required is to lay out the pulley diameters at the desired center distance, as shown in Fig. 11-11. The straight portions B and B' can be scaled, and the lengths of the two arcs A_1 and A_2 can be obtained by measuring the angles θ_1 and θ_2. The arc lengths then are

$$A_1 = \frac{\theta_1}{360} \pi D \quad \text{and} \quad A_2 = \frac{\theta_2}{360} \pi d \qquad (11\text{-}2)$$

The exact point where the belt is tangent to the pulley can be determined graphically by drawing a line perpendicular to the straight portion of the belt and through the center of the pulley.

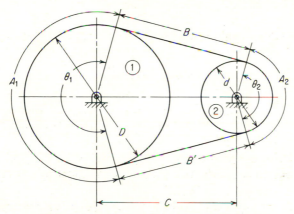

FIG. 11-11.   Center distance and belt length.

The following complete formulas for computing the belt length and center distance are from *The New American Machinists' Handbook:*[1]

$$L = 2C + 1.57(D + d) + \frac{(D - d)^2}{4C} \qquad (11\text{-}3)$$

$$C = \frac{b + \sqrt{b^2 - 32(D - d)^2}}{16} \qquad (11\text{-}4)$$

[1] Rupert LeGrand (ed.), *The New American Machinists' Handbook,* p. 43-34, McGraw-Hill Book Company, Inc., New York, 1955.

where $b = 4L - 6.28(D + d)$
  D = diameter of large pulley
  d = diameter of small pulley
  L = length of belt
  C = center distance

11-3.   Chains

Chains combine the versatility of belts and the positiveness of gears. There are two basic types of chains: the *roller chain* and the *silent chain*. Both types run on toothed wheels, called *sprockets*, and, like belts, may be mounted singly or in gangs.

Figure 11-12 gives a good idea of the versatility and flexibility of a roller chain. It can wrap around a sprocket from either side, can

FIG. 11-12.   Roller-chain installation. (*FMC/Link-Belt.*)

bend easily around small sprockets, and is fairly tolerant of dirty and poorly lubricated conditions.

Figure 11-13 shows a silent chain. These chains are made up of flat steel stampings, which makes it easy to *build up* any width desired. One of their most popular applications is for timing chains in automotive engines, where long life and quiet operation are essential. They are also used extensively for industrial applications requiring the smooth transmission of power at high speeds. Notice how the special links in the center of the chain engage with grooves in the sprockets to prevent the chain from sliding off.

Both roller and silent chains are available in a wide variety of link sizes, and *any* length can be ordered. Standard sprockets are also available in a great variety of widths and diameters. The length and center distance of a particular chain installation can be calculated by Eqs. (11-2), (11-3), and (11-4). The velocity ratio can be calculated by Eq. (11-1), which can be expanded to include numbers of teeth on the sprockets:

$$VR = \frac{\omega_F}{\omega_D} = \frac{D_D}{D_F} = \frac{N_D}{N_F} \qquad (11-5)$$

where N_D, N_F = numbers of teeth on the driver and follower sprockets.

11-4. Gear Trains

The rest of this chapter deals with gear trains of various types. As in other parts of this book, the motions are analyzed with no regard for strength requirements. The methods developed in this chapter can be used to obtain gear data that are satisfactory from a kinematic standpoint, but in an actual design situation the *possible* solutions would be limited by space and torque requirements.

The following general definitions are offered as an aid to understanding the discussions that follow.

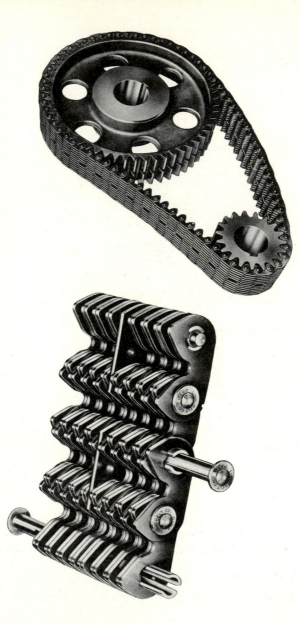

FIG. 11-13.   Silent chain. (*FMC/Link-Belt.*)

GEAR TRAIN

Two or more gears in mesh, transmitting motion (or power) from one shaft to another. Figure 11-14 shows a good example of a gear train.

FIG. 11-14.   Gear train. (*FMC/Link-Belt.*)

SIMPLE TRAIN

A train in which each shaft carries only one gear.

COMPOUND TRAIN

A train in which each shaft, except the first and last, have two gears fastened together on the same shaft to operate as an integral part. The gear drive shown in Fig. 11-7 utilizes a compound train.

SIMPLE-COMPOUND TRAIN

A combination of the two previous types of gear trains.

PINION AND GEAR

With respect to two mating gears, the smaller is called the *pinion*; the larger is called the *gear*.

IDLER GEAR

A gear in a train, other than the first or last gear, that is mounted on a shaft by itself. It acts both as a driver and as a follower. An idler does not affect the speed ratio (train value) of the train; it serves only to *fill up space* and *reverse direction*.

TRANSMISSION OR GEAR BOX

A gear train that has several possible velocity ratios that can be selected by moving one or more of the gears by means of a shifting lever.

The obvious purpose of any gear train is to transmit motion (or power) from one point to another. In addition, however, it is usually necessary to *modify* the motion by increasing or decreasing its speed or by changing its direction.

It would be kinematically possible to obtain almost any velocity ratio with only two gears, but for large speed ratios the size of one of the gears might become prohibitive. Gear trains make it possible to obtain large velocity ratios with a series of small gears. In most cases, a gear train consisting of several pairs of gears can be made more compact than a kinematically equivalent train made up of only two gears. This is possible because the first pairs of gears in a train can usually be made much smaller and lighter owing to the lower torque involved. Notice in Fig. 11-14 that the three sets of gears have progressively larger teeth.

The more important purposes of gear trains can be summarized as follows:

1. To transmit motion (or power) from one point to another point
2. To modify motion by changing its speed or direction
3. To obtain required velocity ratios in a limited space

11-5. Velocity Ratio and Train Value

When two mating gears are used to transmit motion, their velocity ratio, which is the ratio of the output velocity to the input velocity, is of primary interest and is expressed as follows:

$$VR = \frac{\omega_F}{\omega_D} \qquad (11\text{-}6)$$

where ω_D and ω_F are the angular velocities (measured either in radians or in revolutions per unit of time) of the driver and follower. It is conventional for the numerator of the ratio to be the angular velocity of the follower. Also, it is conventional for the velocity ratio to be expressed as a positive number if the driver and follower rotate in the same direction and negative if they rotate in opposite directions.

The velocity ratio may also be expressed in terms of the pitch diameters or the numbers of teeth on the two gears, as well as their angular velocities.

$$VR = \frac{\omega_F}{\omega_D} = \frac{D_D}{D_F} = \frac{N_D}{N_F} \qquad (11\text{-}7)$$

It is important to notice that, in the cases of pitch diameters and numbers of teeth, the data for the followers are shown as the denominators rather than as the numerators. This is because of the inverse relationship that exists between the angular velocities and the diameters of mating gears.

PROOF

In Fig. 11-15 two gears are mounted as shown. Point P is a point that can be considered on either gear (common centro) and, therefore, has the same velocity V on both. The angular velocities of the two gears can then be expressed as

$$V_P = R_D\omega_D$$

and

$$V_P = R_F\omega_F$$

Therefore

$$R_F\omega_F = R_D\omega_D$$

and

$$\frac{\omega_F}{\omega_D} = \frac{R_D}{R_F}$$

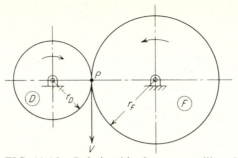

FIG. 11-15. Relationship between radii and angular velocities.

Therefore, *the angular velocities of two mating gears vary inversely as their radii (and, therefore, as their diameters, their circumferences, and their numbers of teeth).*

When a gear train consists of three or more gears, the ratio of its output velocity to its input velocity is referred to as its *train value,* and *the train value is equal to the product of the individual velocity ratios of the drivers and followers making up the train.*

PROOF

Figure 11-16 shows a compound gear train consisting of two pairs of gears whose velocity ratios are $\frac{1}{2}$ and $\frac{1}{3}$, respectively (gears A and B are keyed to the same shaft).

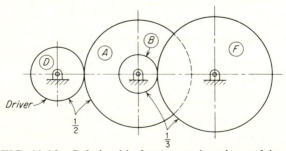

FIG. 11-16. Relationship between train value and individual velocity ratios.

The velocity ratio of the first pair of gears may be expressed as

$$VR_1 = \frac{\omega_A}{\omega_D} = \frac{1}{2}$$

or

$$\omega_A = \frac{1}{2}\omega_D \qquad (a)$$

The velocity ratio of the second pair of gears may be expressed as

$$VR_2 = \frac{\omega_F}{\omega_B} = \frac{1}{3}$$

or

$$\omega_F = \frac{1}{3}\omega_B \qquad (b)$$

Substituting (a) into (b) gives the expression

$$\omega_F = \frac{1}{3}(\frac{1}{2}\omega_D)$$

or

$$\omega_F = \frac{1}{6}\omega_D$$

Therefore,

$$\frac{\omega_F}{\omega_D} = \frac{1}{6}$$

which is the product of $\frac{1}{2}$ and $\frac{1}{3}$.

Therefore, since the train value is equal to the product of the individual velocity ratios and since these ratios can be expressed in terms of angular velocities, pitch diameters, or numbers of teeth, the train value (TV) can be expressed in any of the following ways:

$$TV = \pm \text{ product of individual ratios} \qquad (11\text{-}8)$$

$$= \pm \frac{\text{product of angular velocities of followers}}{\text{product of angular velocities of drivers}} \qquad (11\text{-}9)$$

$$= \pm \frac{\text{product of pitch diameters of drivers}}{\text{product of pitch diameters of followers}} \qquad (11\text{-}10)$$

$$= \pm \frac{\text{product of numbers of teeth on drivers}}{\text{product of numbers of teeth on followers}} \qquad (11\text{-}11)$$

Notice that, in the case of angular velocities, the *followers* are in the numerator, whereas, in the cases of pitch diameters or numbers of teeth, the *drivers* are in the numerator, in accord with Eq. (11-7).

It is conventional to designate the train value as *positive if the driver and follower have the same rotational sense, and negative if they have opposite senses.* The direction (or sense) of the follower of a gear train is often obtained by visually *chasing through* the directions of all of the gears. The follower's direction can be obtained quickly, however, once it is realized that two axles connected by *external* gears rotate in opposite directions. Thus, if the number of axles is *even*, the driver and follower will rotate in opposite directions (as in the case of a pair of gears) and the train value will be negative, whereas, if the number of axles is *odd*, they will rotate in the same direction and the train value will be positive.

The above "axle rule" for follower sense is true whether or not the train contains idlers. Also, Eqs. (11-8) to (11-11) are true, whether or not the train contains idlers. If the train contains an idler, the idler will appear both in the numerator and the denominator and, therefore, cancels itself as far as the absolute value of the train value is concerned—it affects only the sense of the train.

EXAMPLE 11-1. GEAR TRAIN: NUMBERS OF TEETH GIVEN

Figure 11-17 shows a compound gear train consisting of three pairs of gears with the number of teeth on each gear indicated. It is required to find the train value.

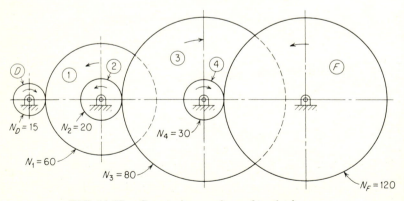

FIG. 11-17.   Gear train: numbers of teeth given.

SOLUTION: Using Eq. (11-11),

$$\text{TV} = \frac{N_D N_2 N_4}{N_1 N_3 N_F} = \frac{(15)(20)(30)}{(60)(80)(120)} = (-)\frac{1}{64}$$

The minus sign indicates that the driver and follower have opposite senses.

EXAMPLE 11-2. GEAR TRAIN: RATIOS GIVEN

Figure 11-18 shows a compound gear train where the individual velocity ratios for the three pairs of gears are given. Find the train value.

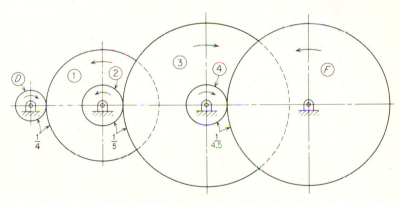

FIG. 11-18.   Gear train: ratios given.

SOLUTION: Using Eq. (11-8),

$$\text{TV} = (\text{VR}_1)(\text{VR}_2)(\text{VR}_3)$$
$$= (\tfrac{1}{4})(\tfrac{1}{5})(\tfrac{1}{4.5}) = (-)\frac{1}{90}$$

EXAMPLE 11-3. GEAR TRAIN: MIXED DATA

Figure 11-19 shows a simple-compound gear train with mixed data. Find the train value.

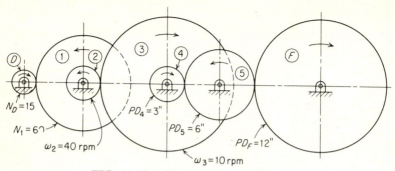

FIG. 11-19. Gear train: mixed data.

SOLUTION

$$\text{TV} = \left(\frac{N_D}{N_1}\right)\left(\frac{\omega_3}{\omega_2}\right)\left(\frac{D_5}{D_4}\right)\left(\frac{D_F}{D_5}\right)$$

$$= (^{15}\!/_{60})(^{10}\!/_{40})(^3\!/_6)(^6\!/_{12}) = (+)\,^1\!/_{64}$$

Notice, in the preceding example, that when the data are mixed, care must be exercised to ensure that the numerator-denominator relationships are in accord with Eqs. (11-7) to (11-11). Also notice that the idler gear 5 in the example does not affect the train value, since it appears as a factor in both the numerator and denominator. Its sole effect is to reverse the sense of the follower.

In the preliminary design of gear trains where several gears are involved, the exact arrangement, sizes, and center distances of the various gears must usually be juggled around to accommodate a given situation. The gears can be strung out in a line or they can be staggered, and idler gears may have to be added to fill up space or to reverse the sense of the output gear. The matter of primary concern is to obtain the desired train value. There are two general approaches, which are illustrated in the following example.

EXAMPLE 11-4. DEVISING COMPOUND GEAR TRAIN

A compound gear train is required that will have a train value of (+) 600 (which means, of course, 600/1). The smallest gear must have no fewer than 12 teeth and the largest no more than 96 teeth.

SOLUTION

1. Determine the maximum velocity ratio for any one of the pairs of gears in the train.

$$VR_{max} = \frac{96}{12} = \frac{8}{1}$$

2. Determine how many pairs of gears are needed. Since the TV is equal to the product of the individual VR's,

$$3 \text{ pairs: } (8)(8)(8) = 512 \text{ (too low)}$$
$$4 \text{ pairs: } (8)(8)(8)(8) = 4,096 \text{ (too high)}$$

It is obvious that four pairs are required.

3. Assuming that three of the pairs will have a velocity ratio of 8, solve for the velocity ratio of the fourth pair.

$$512 \, VR_4 = 600$$

or

$$VR_4 = \frac{600}{512} = \frac{75}{64}$$

Therefore, a suitable train would be

$$TV = \left(\frac{96}{12}\right)\left(\frac{96}{12}\right)\left(\frac{96}{12}\right)\left(\frac{75}{64}\right) = 600$$

where the values in each factor represent numbers of teeth on the four pairs of gears.

It is not always possible to obtain a ratio that can be expressed by numbers falling within the minimum and maximum number of permissible teeth. In this case, it would be necessary to vary one or more of the originally assumed ratios and try again.

ALTERNATIVE SOLUTION

Steps 1 and 2 are the same as above.

3. Once it is determined that four pairs are required, the ideal or theoretical velocity ratio of each pair is obtained as follows:

$$VR_t = \sqrt[4]{TV} = \sqrt[4]{600}$$
$$\log 600 = 2.7782$$
$$(\log 600)/4 = 0.6946$$

then

$$VR_t = 4.95$$

4.   Next, assign a *whole-number* ratio close to this theoretical ratio to three of the pairs, and solve for the velocity ratio of the fourth pair.

$$(5)(5)(5) \, VR_4 = 600$$

$$VR_4 = \frac{600}{125} = \frac{72}{15}$$

Therefore, a suitable train would be

$$TV = \left(\frac{60}{12}\right)\left(\frac{60}{12}\right)\left(\frac{60}{12}\right)\left(\frac{72}{15}\right) = 600$$

Again, it may be necessary to change one or more of the other three velocity ratios to obtain a workable solution.

11-6.   Two Mating Gears

Problems involving two mating gears are concerned primarily with velocity ratio and center distance. The numbers and size (P) of the teeth can usually vary within limits. There are three types of such problems:

Type 1.   Where a specific ratio is required but where the center distance can vary (a trivial, but common, situation).

Type 2.   Where the velocity ratio and center distance are both specified.

Type 3.   Where a decimal velocity ratio and the center distance are specified. In this case it is not always possible to obtain the exact ratio desired. Therefore, either the ratio or the center distance must be varied somewhat.

EXAMPLE 11-5.   TYPE 1

A pair of spur gears is required that will have a velocity ratio of ⅗. The minimum number of teeth on the pinion is to be 12. The center distance is not specified.

SOLUTION

Using Eq. (11-7), the velocity ratio may be expressed as

$$VR = \frac{3}{5} = \frac{N_D}{N_F}$$

This means that for every 3 teeth on the driver there must be 5 teeth on the follower. Since the minimum number of teeth on the smaller gear must be 12, both parts of the ratio must be multiplied by some number that will increase 3 to 12 (or more). Such a number is 4; so

$$N_D = (3)(4) = 12 \text{ teeth}$$
$$N_F = (5)(4) = 20 \text{ teeth}$$

Now, since the size of the teeth was not specified, *any* diametral pitch may be selected, but the larger the tooth size, the greater will be the center distance. Although any diametral pitch may arbitrarily be chosen as far as a satisfactory kinematic solution is concerned (strength is not considered here), it is helpful to choose one, if possible, that will divide evenly into the number of teeth on each gear so that the resulting pitch diameters and center distance will be *convenient* values.

Selecting $P = 4$ results in the following pitch diameters and center distance:

$$D_D = \frac{N_D}{P} = \frac{12}{4} = 3 \text{ in.}$$

$$D_F = \frac{N_F}{P} = \frac{20}{4} = 5 \text{ in.}$$

$$C = \frac{D_D + D_F}{2} = \frac{3 + 5}{2} = 4 \text{ in.}$$

It must be emphasized that the solution in the preceding example is just one of many possible solutions. If there were some other requirements to meet, such as space or strength, other more appropriate solutions could easily be found.

EXAMPLE 11-6.   TYPE 2

Figure 11-20 shows the requirements for two mating gears. Their center distance must be $6\frac{1}{2}$ in. and the follower must rotate at 200 rpm when the driver rotates at 500 rpm. Find the pitch diameters, numbers of teeth, and diametral pitch for two gears to accomplish this.

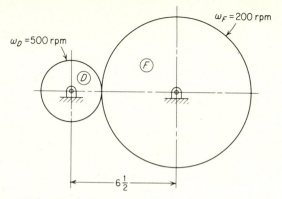

FIG. 11-20. Two mating gears: ratio and center distance specified (type 2).

SOLUTION

1. Using Eq. (11-7), the velocity ratio may be expressed as

$$VR = \frac{\omega_F}{\omega_D} = {}^{200}\!/_{500} = {}^{2}\!/_{5}$$

or as

$$VR = \frac{D_D}{D_F} = {}^{2}\!/_{5}$$

2. A rearrangement of Eq. (10-16) produces the expression

$$D_D + D_F = 2C$$

or

$$D_D + D_F = 13$$

3. Then, from the second expression in step 1, it is clear that the pitch diameters of the two gears must be in the ratio of 2:5 and the total parts of the ratio are $2 + 5 = 7$; so

$$D_D = ({}^{2}\!/_{7})(13) = {}^{26}\!/_{7} \text{ (in.)}$$

and

$$D_F = ({}^{5}\!/_{7})(13) = {}^{65}\!/_{7} \text{ (in.)}$$

4. Since $D = N/P$ [Eq. (10-1)], a possible solution is obtained by letting

$$N_D = 26 \text{ teeth}$$

$$N_F = 65 \text{ teeth}$$

and $\quad\quad\quad\quad\quad P = 7$

then $\quad\quad\quad\quad D_D = \; ^{26}\!/_7 = 3^5\!/_7$ in.

and $\quad\quad\quad\quad D_F = \; ^{65}\!/_7 = 9^2\!/_7$ in.

If the diametral pitch (tooth size) must be larger or smaller, other solutions can be obtained by multiplying or dividing both the N's and the P by some common number. For example, if they were multiplied by 2, another solution would be obtained; this would be $N_D = 52$, $N_F = 130$, and $P = 14$. Care must be used in selecting a number to multiply or divide by to ensure that the N's remain whole numbers, and preferably the pitch also.

ALTERNATIVE SOLUTION

Steps 1 and 2 are the same as above.

3. Solve the second expression in step 1 for D_F.

$$D_F = (^5\!/_2)D_D$$

4. Substitute this expression for D_F into the expression shown in step 2, and solve for D_D.

$$D_D + (^5\!/_2)D_D = 13$$

or $\quad\quad\quad\quad\quad\quad\quad D_D = \; ^{26}\!/_7$ (in.)

5. Substitute the value just obtained for D_D back into the expression in step 2, and solve for D_F.

$$^{26}\!/_7 + D_F = 13$$

or $\quad\quad\quad\quad\quad\quad\quad D_F = \; ^{65}\!/_7$ (in.)

6. Again, since $D = N/P$ [Eq. (10-1)], a possible solution is obtained by letting

$$N_D = 26 \text{ teeth}$$
$$N_F = 65 \text{ teeth}$$

and $\quad\quad\quad\quad\quad P = 7$

then $\quad\quad\quad\quad D_D = \; ^{26}\!/_7 = 3^5\!/_7$ in.

$$D_F = \; ^{65}\!/_7 = 9^2\!/_7 \text{ in.}$$

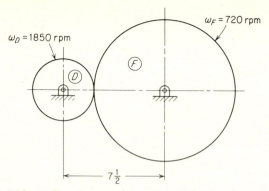

FIG. 11-21.   Two mating gears: decimal ratio and center distance specified (type 3).

EXAMPLE 11-7.   TYPE 3

Figure 11-21 shows the requirements for two mating gears. Their center distance must be $7\frac{1}{2}$ in., and it is desired that the follower rotate as close to 720 rpm as possible when the driver rotates at 1,850 rpm. Find the pitch diameters, numbers of teeth, and diametral pitch for two gears that will accomplish this. The pinion must have a minimum of 12 teeth.

SOLUTION

1. Using Eq. (11-7), the velocity ratio may be expressed as

    $$\text{VR} = \frac{\omega_F}{\omega_D} = \frac{720}{1,850} = \frac{1}{2.565}$$

    also

    $$\text{VR} = \frac{D_D}{D_F} = \frac{1}{2.565}$$

2. A rearrangement of Eq. (10-16) produces the expression

    $$D_D + D_F = 2C$$

    or

    $$D_D + D_F = 15$$

3. Then, from the second expression in step 1, it is clear that the pitch diameters should be in the ratio of $1:2.565$ and the total parts of the ratio are $1 + 2.565 = 3.565$; so

$$D_D = \left(\frac{1}{3.565}\right)(15) = \frac{15}{3.565} \text{ (in.)}$$

and
$$D_F = \left(\frac{2.565}{3.565}\right)(15) = \frac{38.48}{3.565} \text{ (in.)}$$

4. Since $D = N/P$, a theoretical "solution" is obtained by letting

$$N_D = 15$$
$$N_F = 38.48$$
and
$$P = 3.565$$

Such a "solution" is not practical for it involves fractions of teeth on a gear (impossible) and it involves a decimal diametral pitch (most undesirable).

5. Next, then, multiply all of these values by a number that will make P an integer. Since N_D and N_F already exceed the minimum of 12, raising the value of P to 4 will be tried. Such a number is (4/3.565). This results in

$$N_D = (15)\left(\frac{4}{3.565}\right) = 16.8 \approx 17 \text{ teeth}$$

$$N_F = (38.48)\left(\frac{4}{3.565}\right) = 43.2 \approx 43 \text{ teeth}$$

$$P = (3.565)\left(\frac{4}{3.565}\right) = 4$$

then
$$D_D = \frac{N_D}{P} = \frac{17}{4} = 4\frac{1}{4} \text{ in.}$$

$$D_F = \frac{N_F}{P} = \frac{43}{4} = 10\frac{3}{4} \text{ in.}$$

To check these values:

$$C = \frac{N_D + N_F}{2P} = \frac{17 + 43}{(2)(4)} = 7\frac{1}{2} \text{ in.}$$

or
$$C = \frac{D_D + D_F}{2} = \frac{4\frac{1}{4} + 10\frac{3}{4}}{2} = 7\frac{1}{2} \text{ in.}$$

and
$$VR = \frac{N_D}{N_F} = \frac{17}{43} = \frac{1}{2.53}$$

and since
$$VR = \frac{\omega_F}{\omega_D} = \frac{1}{2.53}$$

the resulting rpm of the follower is

$$\frac{\omega_F}{1,850} = \frac{1}{2.53}$$

or
$$\omega_F = 732 \text{ rpm}$$

If the resulting rpm of the follower is not close enough, a closer value could be obtained by going back to step 5 and multiplying by a number that will result in a higher value of P (smaller teeth) and then rounding off the results to an integral number of teeth, as before.

ALTERNATIVE SOLUTION

Steps 1 and 2 are the same as above.

3. Solve the second expression in step 1 for D_F.

$$D_F = 2.565 D_D$$

4. Substitute this expression for D_F into the expression shown in step 2, and solve for D_D.

$$D_D + (2.565)D_D = 15$$

or
$$D_D = \frac{15}{3.565} \text{ (in.)}$$

5. Substitute the value just obtained for D_D back into the expression in step 2, and solve for D_F.

$$\frac{15}{3.565} + D_F = 15$$

or
$$D_F = \frac{53.48}{3.565}$$

The above results are the same as those obtained in step 3 of the original solution. The alternative solution would now be the same as that given in step 4 of the original solution.

11-7. Reverted Gear Trains

A reverted gear train is a train in which the driver and follower are coaxial, that is, turn about the same axis. Typical uses include automotive transmissions, airplane-propeller drives, and machine-tool spindle drives.

In designing reverted gear trains, two conditions must be met: (1) the overall train value must be correct, which means that the velocity ratios of the two pairs of gears must be *factors* of the train value; and (2) the center distances of the two pairs of gears must be the same. The methods employed in designing such gears are illustrated in the following two examples.

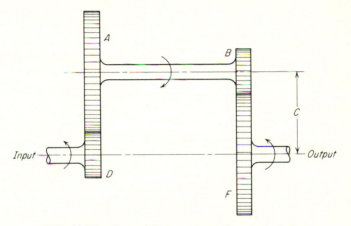

FIG. 11-22. Reverted gears: same diametral pitch.

Example 11-8. Reverted Gears: Same Diametral Pitch

A set of reverted gears, as shown in Fig. 11-22, is required that will have a train value of 1:20. The minimum number of teeth on any gear is to be 12, and the diametral pitch for both pairs is to be the same. Find the numbers of teeth for a set of gears to accomplish this.

SOLUTION

1. Divide the train value into two factors as nearly equal as possible. In this case, the ideal factor would be $\sqrt{20}$, but, since decimal ratios

are to be avoided if possible, it is better to use the factors 4 and 5; so the ratios of 1:4 and 1:5 are assigned to the two pairs.

$$\text{Ratio of first pair: } VR_1 = \frac{N_D}{N_A} = \frac{1}{4}$$

$$\text{Ratio of second pair: } VR_2 = \frac{N_B}{N_F} = \frac{1}{5}$$

2. The center distance may be expressed as

$$C_1 = \frac{N_D + N_A}{2P} \quad \text{and} \quad C_2 = \frac{N_B + N_F}{2P}$$

and since the center distances must be the same,

$$\frac{N_D + N_A}{2P} = \frac{N_B + N_F}{2P}$$

and since, in this case, their diametral pitches are to be the same

$$N_D + N_A = N_B + N_F = M$$

where M is called a *multiple number*, which is selected as follows:

The total number of teeth of the first pair ($N_D + N_A$) must divide into the ratio of 1:4, and the total number of teeth on the second pair ($N_B + N_F$) must divide into the ratio of 1:5. Therefore, the multiple number M must be divisible by both (1 + 4) and (1 + 5). Such a number would be (5)(6) = 30.

3. Determine the numbers of teeth for the first pair as follows:

Since $\qquad\qquad N_D + N_A = 30$

and $\qquad\qquad \dfrac{N_D}{N_A} = \dfrac{1}{4}$

then $\qquad\qquad N_D = (\tfrac{1}{5})(30) = 6$ teeth

and $\qquad\qquad N_A = (\tfrac{4}{5})(30) = 24$ teeth

4. Similarly, determine the numbers of teeth for the second pair.

Since $\qquad\qquad N_B + N_F = 30$

and $\qquad\qquad \dfrac{N_B}{N_F} = \dfrac{1}{5}$

then
$$N_B = (\tfrac{1}{6})(30) = 5 \text{ teeth}$$
and
$$N_F = (\tfrac{5}{6})(30) = 25 \text{ teeth}$$

But the numbers of teeth for gears B and D are below the minimum of 12. Therefore, all the above values must be multiplied by some number that will result in the smallest gear having at least 12 teeth. Such a number is 3.

Therefore,
$$N_D = (6)(3) = 18 \text{ teeth}$$
$$N_A = (24)(3) = 72 \text{ teeth}$$
$$N_B = (5)(3) = 15 \text{ teeth}$$
$$N_F = (25)(3) = 75 \text{ teeth}$$

Now, any diametral pitch may be selected, but the resulting pitch diameters and center distance will vary accordingly.

ALTERNATIVE SOLUTION

1. From step 1 in the original solution
$$\frac{N_D}{N_A} = \tfrac{1}{4}$$
and
$$\frac{N_B}{N_F} = \tfrac{1}{5}$$

2. From step 2 in the original solution
$$N_D + N_A = N_B + N_F$$

3. Solve the first expression from step 1 for N_A and solve the second expression from step 1 for N_F.
$$N_A = 4\,N_D$$
and
$$N_F = 5\,N_B$$

4. Substitute these expressions for N_A and N_F into the equation from step 2.
$$N_D + 4N_D = N_B + 5N_B$$
then
$$5N_D = 6N_B$$
or
$$\frac{N_D}{N_B} = \tfrac{6}{5}$$

5. A solution can be obtained by letting $N_D = 6$ and $N_B = 5$ and using
    the expressions from step 3 to solve for N_A and N_F.

Let $N_D = 6$ teeth

then $N_A = 4 N_D = (4)(6) = 24$ teeth

Let $N_B = 5$ teeth

then $N_F = 5 N_B = (5)(5) = 25$ teeth

These results are the same as those obtained in the original example,
and, again, it is necessary to multiply all of these values by 3 to satisfy
the minimum of 12 teeth.

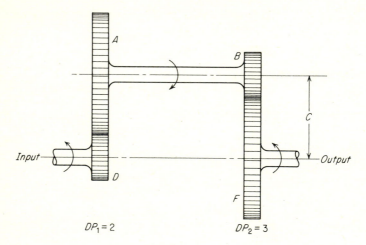

FIG. 11-23.   Reverted gears: different diametral pitches.

EXAMPLE 11-9.   REVERTED GEARS: DIFFERENT DIAMETRAL PITCHES

A set of reverted gears, as shown in Fig. 11-23, is required that will have
a train value of $1:20$. The diametral pitch of the first pair of gears is to
be 2, and the diametral pitch of the second pair is to be 3. The minimum
number of teeth on any gear is to be 12. Find the numbers of teeth for a
set of gears to accomplish this.

SOLUTION

1. Again, as in the preceding example, assign the ratios of $1:4$ and $1:5$.

$$\text{Ratio of first pair: } VR_1 = \frac{N_D}{N_A} = \frac{1}{4}$$

$$\text{Ratio of second pair: } VR_2 = \frac{N_B}{N_F} = \frac{1}{5}$$

2. The center distance may be expressed as

$$C = \frac{N_D + N_A}{2P_1} = \frac{N_B + N_F}{2P_2}$$

Then, cancelling 2's, cross-multiplying, and equating the two expressions to an M (multiple number),

$$P_2(N_D + N_F) = P_1(N_B + N_F) = M$$

The value of M must be such that it is divisible by both $(1 + 4)$ and $(1 + 5)$ *after* it has been divided by *either* of the two diametral pitches. A number that will certainly fulfill this requirement is the product of all the numbers involved (or their least common multiple). Such a number is $(1 + 4)(1 + 5)(2)(3) = 180$. Therefore, let $M = 180$.

3. Determine the numbers of teeth for the first pair of gears.

Since $\quad\quad\quad P_2(N_D + N_F) = 180$

and $\quad\quad\quad\quad\quad\quad P_2 = 3$

then $\quad\quad\quad\quad\quad N_D + N_F = 60$

and since $\quad\quad\quad\quad\quad \dfrac{N_D}{N_A} = \dfrac{1}{4}$

then $\quad\quad\quad\quad N_D = (\frac{1}{5})(60) = 12$ teeth

and $\quad\quad\quad\quad N_A = (\frac{4}{5})(60) = 48$ teeth

4. Determine the numbers of teeth for the second pair of gears.

Since $\quad\quad\quad P_1(N_B + N_F) = 180$

and $\quad\quad\quad\quad\quad\quad P_1 = 2$

then $\quad\quad\quad\quad\quad N_B + N_F = 90$

and since $\qquad \dfrac{N_B}{N_F} = \frac{1}{5}$

then $\qquad N_B = (\frac{1}{6})(90) = 15$ teeth

and $\qquad N_F = (\frac{5}{6})(90) = 75$ teeth

To check these values:

$$\text{TV} = \left(\frac{N_D}{N_A}\right)\left(\frac{N_B}{N_F}\right) = \left(\frac{12}{48}\right)\left(\frac{15}{75}\right) = \frac{1}{20}$$

$$C_1 = \frac{N_D + N_A}{2P_1} = \frac{(12 + 48)}{(2)(2)} = 15 \text{ in.}$$

$$C_2 = \frac{N_B + N_F}{2P_2} = \frac{(15 + 75)}{(2)(3)} = 15 \text{ in.}$$

ALTERNATIVE SOLUTION

1. From step 1 in the original solution

$$\frac{N_D}{N_A} = \frac{1}{4}$$

and $\qquad \dfrac{N_B}{N_F} = \frac{1}{5}$

2. From step 2 in the original solution

$$P_2(N_D + N_A) = P_1(N_B + N_F)$$

3. Solve the first expression from step 1 for N_A and solve the second expression from step 1 for N_F.

$$N_A = 4N_D$$

and $\qquad N_F = 5N_B$

4. Substitute these expressions for N_A and N_F into the equation from step 2.

$$P_2(N_D + 4N_D) = P_1(N_B + 5N_B)$$

then $\qquad (3)(5N_D) = (2)(6N_B)$

$$15N_D = 12N_B$$

or $\qquad \dfrac{N_D}{N_B} = \frac{12}{15}$

5. A solution can be obtained by letting $N_D = 12$ and $N_B = 15$ and using the expressions from step 3 to solve for N_A and N_F.

Let	$N_D = 12$ teeth
then	$N_A = 4N_D = (4)(12) = 48$ teeth
Let	$N_B = 15$ teeth
then	$N_F = 5N_B = (5)(15) = 75$ teeth

These results are the same as those obtained in the original solution.

11-8.   Epicyclic Gear Trains (Planetary Gears)

In the gear trains discussed to this point, all the gears rotated about fixed axes, the frame being the fixed link in the mechanism. In epicyclic gear trains, however, the axes of certain gears are in motion, and one of the gears becomes a part of the fixed link, or frame. An epicyclic gear train may then be described as a train in which one gear is fixed (the sun gear) and the other gears (planet gears), carried by a revolving arm (or carrier), rotate not only about their own centers, but their centers rotate about the fixed gear, as illustrated in Fig. 11-24a. Epicyclic gears are so named because the paths of the planet-gear teeth are epicycloids, as shown in Fig. 11-24b.

Figure 11-25 shows an epicyclic gear train with four planet gears operating about a small sun gear. The planet gears, in turn, drive an annular ring gear.

The angular velocities of gears in conventional gear trains can be determined quite simply by using the ratios of the numbers of teeth on the various gears. For example, in Fig. 11-26a, if the angular velocity of gear B were known, the angular velocity of gear C could be obtained, for with every revolution of gear B, gear C makes $^{40}/_{10}$, or 4 revolutions.

In Fig. 11-26b, however, it is not quite so obvious how many revolutions B makes for each revolution of arm A. To assist in following C's motion, a mark is made on C where it contacts B, as

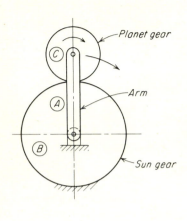

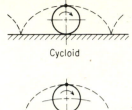

Cycloid

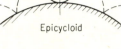

Epicycloid

Hypocycloid

(a)   (b)

FIG. 11-24.   Epicyclic (planetary) gears.

FIG. 11-25.   Epicyclic gear train. (*The Cincinnati Gear Company.*)

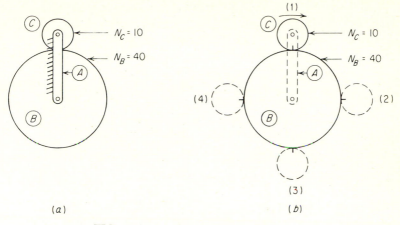

FIG. 11-26. Epicyclic-gear analysis.

shown in position 1. If arm A is rotated clockwise until all of C's teeth have touched gear B, or until the mark is again contacting B, arm A will have made $\frac{1}{4}$ revolution and gear C will be as shown at position 2, but it is evident that gear C has made $1\frac{1}{4}$ revolutions about its own axis. To arrive at position 3, gear C will have made a total of $2\frac{1}{2}$ revolutions, and by the time it arrives back at position 1, gear C will have made 5 revolutions, although the gear-tooth ratio is 4. In other words, the angular-motion or angular velocity ratio for epicyclic gears is not the same as their gear-tooth ratios, as is the case in conventional gear trains.

In analyzing the motion of epicyclic gear trains, then, it is helpful to introduce two imaginary steps. The first step sets the stage for the second step, which treats the gear train as a conventional one where all gears rotate about fixed axes.

TWO-STEP ANALYSIS OF EPICYCLIC GEAR TRAINS

The following two steps are the same, no matter which member in the train is the driver and which is the follower.

Step 1.   Imagine that the entire gear train is locked together as an integral part; then rotate the entire unit one revolution clockwise.[1] As a result, every member in the train will rotate $+1$ revolution.

Step 2.   Now imagine that the gears are again free to turn, and while holding the arm (carrier) stationary, rotate the fixed gear one revolution counterclockwise (-1), back to its original position.

If the movement of each member is recorded for these two steps, it will become evident that the arm has a net motion of one revolution clockwise, the fixed gear has a net motion of zero, and all of the other members' motions will be available for comparison with each other.

The second step of the above procedure makes sense because, in effect, the fixed gear is being backed up to its original position from where it never should have moved. More importantly, in this second step, the gear train is operating like a conventional gear train, and the motions of the various members are easy to determine.

The following examples will serve to illustrate the two-step process for analyzing epicyclic gear trains.

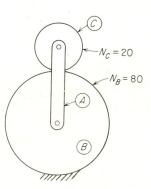

FIG. 11-27.   Epicyclic gear train.

---

[1] It is conventional in gear analysis to think of clockwise as positive, as viewed from the front, top, or right side. Any convention is satisfactory, however, if applied consistently.

EXAMPLE 11-10

Figure 11-27 shows the simplest form of epicyclic gear train. The fixed or sun gear has 80 teeth, and the planet gear has 20 teeth. Find the train value.

SOLUTION

Step	Driver Arm A	Fixed Gear B	Follower Gear C
1	+1	+1	+1
2	0	−1	$+\,^{80}\!/_{20} = +4$
Total	+1	0	+5

EXPLANATION

Step 1.   The entire assembly is imagined to be locked together and rotated one revolution clockwise. Thus, each element rotates +1 revolution.

Step 2.   The driver, arm A, is held stationary, and the *fixed* gear B is rotated one revolution counterclockwise (-1) to its original position. It is evident that, when gear B is rotated one revolution in a negative direction, gear C must rotate N_B/N_C revolutions in a positive direction, or

$$\frac{N_B}{N_C} = \frac{80}{20} = (+)4$$

Since the driver arm A is held stationary during the second step, its motion is zero. If these results are entered in the tabulation and the two steps are totaled, it is evident that the follower makes five revolutions for every revolution of the driver and in the same direction. Therefore,

$$TV = \frac{\omega_F}{\omega_D} = \frac{+5}{+1} = +5$$

EXAMPLE 11-11

Figure 11-28 shows an epicyclic gear train utilizing an *annular* gear as a sun gear. The sun gear has 80 teeth, and the follower gear has 20 teeth. Find the train value.

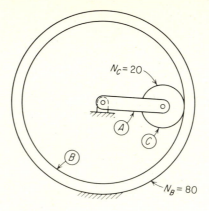

FIG. 11-28.   Epicyclic gear train: with annular gear.

SOLUTION

Step	Driver Arm A	Fixed Gear B	Follower Gear C
1	$+1$	$+1$	$+1$
2	0	-1	$-{}^{80}/_{20} = -4$
Total	$+1$	0	-3

Therefore,

$$\text{TV} = \frac{-3}{+1} = -3$$

EXAMPLE 11-12

Figure 11-29 shows an epicyclic gear train utilizing an idler gear. The sun gear has 80 teeth, the idler gear C has 16, and the follower gear D has 20 teeth. Find the train value.

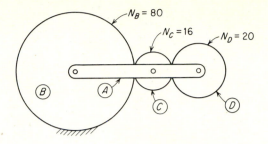

FIG. 11-29. Epicyclic gear train: with idler.

SOLUTION

Step	Driver Arm A	Fixed Gear B	Idler Gear C	Follower Gear D
1	$+1$	$+1$	$+1$	$+1$
2	0	-1	$+\,^{80}/_{16} = +5$	$-\,^{80}/_{20} = -4$
Total	$+1$	0	$+6$	-3

Therefore,

$$\text{TV} = \frac{-3}{+1} = -3$$

Notice that the idler gear in this case accomplished the same results as the annular gear in the preceding example. Notice also that the tabulation for the idler gear C is unnecessary as far as the train value is concerned.

Example 11-13

Figure 11-30 shows a *reverted* epicyclic gear train, where arm A is the driver, gear E is the follower, and gear B is the fixed sun gear. Gears C and D are the planetary gears. The numbers of teeth are as follows: $N_B = 48$, $N_C = 12$, $N_D = 15$, and $N_E = 45$. Find the train value.

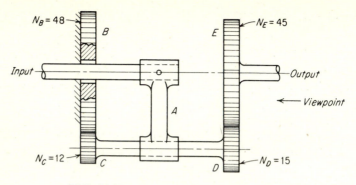

FIG. 11-30. Reverted epicyclic gear train.

SOLUTION

Step	Driver Arm A	Fixed Gear B	Gears C and D	Follower Gear E
1	$+1$	$+1$	$+1$	$+1$
2	0	-1	$+^{48}/_{12} = +4$	$(-)^{48}/_{12} \times {}^{15}/_{45} = -^4/_3$
Total	$+1$	0	$+5$	$-^1/_3$

Therefore,

$$TV = \frac{-^1/_3}{+1} = -\frac{1}{3}$$

If the input angular velocity were 100 rpm, the output rpm could be found as follows:

Since

$$TV = \frac{\omega_F}{\omega_D}$$

then

$$\frac{\omega_F}{100} = (-)\frac{1}{3}$$

or

$$\omega_F = (-)\frac{(100)}{3} = (-)33.3 \text{ rpm}$$

Problems

BELTS AND CHAINS

11-1.   A motor drives a compressor by means of a V belt. If the motor speed is 1,750 rpm and if the motor and compressor pulleys are 2.5 and 8 in. in diameter, respectively, what is the speed of the compressor?

11-2.   A shaft is to rotate at 1,000 rpm. If it is to be driven through a belt or chain by a motor that operates at 1,750 rpm, and if the shaft pulley (or sprocket) is 7.5 in. in diameter, what diameter motor pulley (or sprocket) is required?

11-3.   Two sheaves (pulleys) have diameters of 12 and 6 in., respectively, and the distance between their centers is 11.5 in. Find the length of belt required to connect them.

11-4.   Two sheaves have diameters of 12 and 24 in., respectively, and a belt is available that is 104 in. long. To utilize this belt, what must the center distance be?

11-5.   A chain is to be used to connect a motor to a machine. The motor sprocket has 12 teeth, and the motor operates at 1,750 rpm. How many teeth should the machine sprocket have to operate at 840 rpm?

11-6.   If a 14-in.-diameter sheave, rotating at 250 rpm, drives a 10-in.-diameter sheave, what is the speed of the driven sheave?

11-7.   Two sheaves 64 in. apart are to be connected with a belt 172 in. long. If one of the sheaves is 18 in. in diameter, what must be the diameter of the other sheave?

COMPOUND GEAR TRAINS

11-8.   A compound gear train has three pairs of gears and a train value of 1:60. If the velocity ratios of two of the pairs are 1:5 and 1:3, what is the ratio of the third pair?

11-9.   Find the train value of the gear train shown in Fig. 11-31.

11-10.   Devise a compound gear train with a train value of 900. No gear should have fewer than 15 teeth or more than 90 teeth.

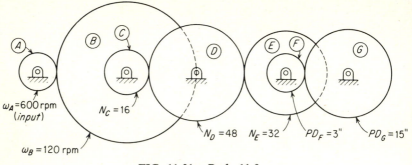

FIG. 11-31.   Prob. 11-9.

Make a sketch and show the number of teeth for each gear. If necessary, add an idler gear to achieve a positive TV.

11-11.   Devise a compound gear train with a train value of $(-)$ 1:400. No gear should have fewer than 12 or more than 72 teeth. Make a sketch and show the number of teeth for each gear. If necessary, add an idler gear to achieve a negative TV.

TWO MATING GEARS

11-12.   If a driving gear with a 3-in. pitch diameter, rotating at 1,320 rpm, meshes with a gear having a pitch diameter of 22 in., what is the speed of the driven gear?

11-13 (*type 1 mating-gear problem*).   Two mating gears are required that will have a velocity ratio of 3:8. Specify the numbers of teeth, pitch diameters, diametral pitch, and center distance for two such gears.

11-14 (*type 2 mating-gear problem*).   Two gears are to be mounted 5.25 in. apart and have a velocity ratio of $\frac{1}{6}$. Find the data (N, P, and D) for two gears that will accomplish this. Neither gear is to have fewer than 15 teeth.

11-15 (*type 2 mating-gear problem*).   Two gears are to be mounted 12 in. on centers and have a velocity ratio of 3. Find the data (N, P, and D) for two gears that will accomplish this. Neither gear should have fewer than 15 teeth and the minimum gear tooth size is $4P$.

11-16 (*type 3 mating-gear problem*).   Two parallel shafts are 7 in. apart, and one shaft operates at 315 rpm. It is desired to drive the other shaft as close to 200 rpm as possible. Determine the data (*N*, *P*, and *D*) for two gears that will accomplish this. The pinion (smaller gear) should have a minimum of 20 teeth. After obtaining the gear data, check the resulting *actual* rpm of the second shaft to see if it is within ± 5 rpm of the desired value.

REVERTED GEARS

11-17.   The reverted gear train in Fig. 11-32 is to have a train value of 1:6, and the diametral pitch can be the same for both pairs of gears. Find the data for four gears that will accomplish this. No gear should have fewer than 12 teeth.

11-18.   The reverted gear train in Fig. 11-32 is to have a train value of 1:15. The diametral pitch of the first pair is to be 6, and the diametral pitch of the second pair is to be 4. No gear is to have fewer than 12 teeth. Find the data for four gears that will accomplish this.

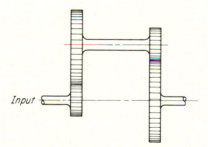

FIG. 11-32.   Probs. 11-17, 11-18, and 11-19.

11-19.   Same as preceding problem except that the TV is 1:30.

EPICYCLIC GEARS

11-20.   Figure 11-33 shows an epicyclic gear train where arm A is the driver and gear C is the follower. Find the train value.

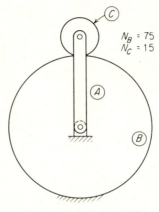

$$N_B = 75$$
$$N_C = 15$$

FIG. 11-33.   Prob. 11-20.

11-21.   Figure 11-34 shows an epicyclic gear train where arm A is the driver and gear B is the follower. Find the train value.

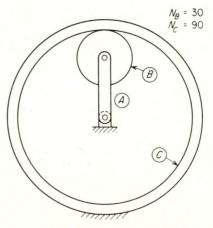

$$N_B = 30$$
$$N_C = 90$$

FIG. 11-34.   Prob. 11-21.

11-22. Figure 11-35 shows an epicyclic gear train where the arm *A* is the driver and the annular gear *D* is the follower. (*a*) Find the train value. (*b*) If the input is 100 rpm, what is the output?

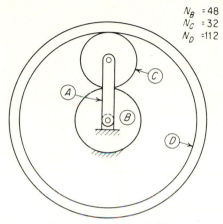

$$N_B = 48$$
$$N_C = 32$$
$$N_D = 112$$

FIG. 11-35.   Prob. 11-22.

11-23. Figure 11-36 shows an epicyclic gear train where gear *B* is the driver and arm *A* is the follower. (*a*) Find the train value. (*b*) If the input is 500 rpm, what is the output?

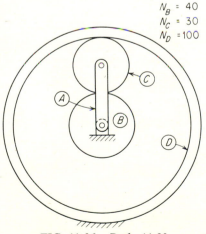

$$N_B = 40$$
$$N_C = 30$$
$$N_D = 100$$

FIG. 11-36.   Prob. 11-23.

11-24. Figure 11-37 shows an epicyclic gear train where the arm *A* is the driver and gear *D* is the follower. (*a*) Find the train value. (*b*) If the input is 120 rpm, what is the output?

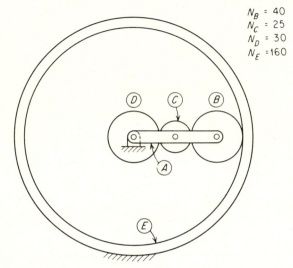

$N_B = 40$
$N_C = 25$
$N_D = 30$
$N_E = 160$

FIG. 11-37. Probs. 11-24 and 11-25.

11-25. Same as preceding problem but consider gear *D* fixed and gear *E* the follower. Find the train value.

11-26. Figure 11-38 shows a *reverted* epicyclic gear train where arm *A* is the driver and gear *E* is the follower. Find (*a*) the train value and (*b*) the velocity ratio between gears *D* and *E*.

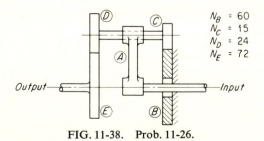

$N_B = 60$
$N_C = 15$
$N_D = 24$
$N_E = 72$

FIG. 11-38. Prob. 11-26.

11-27.   Figure 11-39 shows a *reverted* epicyclic gear train utilizing *bevel* gears. The driver is arm *A* and the follower is gear *E*. Gears *C* and *D* are both fixed to their shaft, and gear *B* is stationary. (*a*) Find the train value. (*b*) If the input is 120 rpm, what is the output?

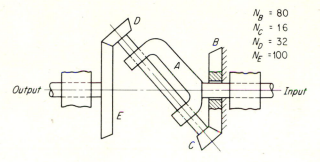

$$N_B = 80$$
$$N_C = 16$$
$$N_D = 32$$
$$N_E = 100$$

FIG. 11-39.   Prob. 11-27.

miscellan-
eous
mechanisms

12-1. Introduction

Over the past several hundred years, there have been countless mechanisms and devices conceived to perform specific mechanical functions. From this endless struggle to develop new and better mechanisms there have evolved many standard devices that can be purchased to handle recurring design problems.

Machine elements such as belts, chains, pulleys (sheaves), sprockets, gears, speed reducers, transmissions, variable-speed drives, flexible couplings, universal joints, and clutches are available commercially from stock. These elements, or components, together with other standard hardware items such as fasteners (nuts, bolts, rivets, etc.), bearings, springs, air and hydraulic actuators, valves, solenoids, wheels, motors, and engines, give the present-day engineer a tremendous head start in solving specific design problems. Any engineer who ignores these many items and designs his equipment "from the ground up" is in effect *reinventing the wheel*, and is wasting his company's time and money. If an engineer is to be an effective designer of equipment, he must keep current on the many standard devices at his disposal.

The object of this chapter is to present a glimpse of some of the standard components that are available, to briefly explain their uses, and finally to illustrate several of the classical methods of producing intermittent motion.[1]

12-2.   Speed Reducers

Speed reducers are compact, self-contained gear trains that may be interposed between an engine or a motor and a machine. They are available in numerous shapes, sizes, shaft arrangements, torque capacities, and velocity ratios. Figure 12-1 shows a speed reducer

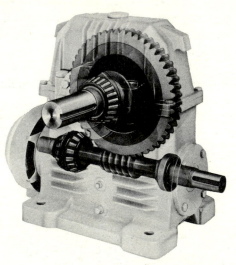

FIG. 12-1.   Speed reducer. (*FMC/Link-Belt.*)

utilizing a worm-gear set. Others are available utilizing other types of gears, such as the helical-gear speed reducer shown in Fig. 11-14. Velocity ratios are available from near 1:1 to 1:10,000, shaft arrangements in straight-through, offset, or right-angle, and power capacities from fractional to several hundred horsepower.

[1] For a thorough treatment of various mechanisms for generating various kinds of motion, see S. B. Tuttle, *Mechanisms for Engineering Design*, John Wiley & Sons, Inc., New York, 1967.

Speed reducers are also available with integral motors, integral fluid couplings, etc. When combined with motors they are often referred to as *gearmotors* or *motoreducers*.

12-3.   Variable-speed Drives

Machines or equipment must often be capable of operating at various speeds to meet varying conditions. For example, the speeds of machine tools must be variable to control cutting speeds, and the speeds of assembly-line conveyors must be adjustable to meet production-timing requirements. Many machines are built with integral change-gear boxes to provide various speed *steps*, but it is becoming increasingly popular to accomplish speed changes with a *stepless* infinitely-variable-speed drive. In most cases, it is more economical for equipment manufacturers to incorporate a standard, proved, variable-speed drive into their machines than it would be to experiment with a design of their own.

The PIV (positive infinitely variable) variable-speed drive shown in Fig. 12-2 is a device with a positive chain drive from the input

FIG. 12-2.   PIV variable-speed drive. (*FMC/Link-Belt.*)

shaft to the output shaft. This is accomplished by a self-tooth-forming chain running on the grooved faces of opposing conical wheels. The effective diameters of these wheels can be altered during operation to change their velocity ratio.

The variable-speed drive shown in Fig. 12-3a is a light-duty, inexpensive, infinitely-variable drive that uses standard V belts. Moving the speed-selector lever changes the sheave pitch diameters and the velocity ratio. The sheave on one side becomes narrower, thus increasing its effective diameter as the sheave on the other side becomes wider, decreasing its effective diameter. Figure 12-3b shows the construction of the sheaves.

(a)

(b)

FIG. 12-3.   Variable-speed drive: V belt. (*Lovejoy Flexible Coupling Company.*)

Figure 12-4 shows an intermediate-duty, variable-speed drive utilizing a spring-loaded split sheave whose effective diameter changes as the motor position is changed.

Figure 11-5 (Chap. 11) shows two heavy-duty variable-speed drives utilizing a special high-strength wide belt. These drives are available in ratings as high as 25 horsepower.

FIG. 12-4.   Variable-speed sheave. (*Lovejoy Flexible Coupling Company.*)

Figure 12-5 shows a light-duty mechanical variable-speed drive which gives stepless variable speed from zero to maximum by changing the distance that four or more one-way clutches rotate the output shaft when they move back and forth successively. These devices are available in ratings from 12 to 100 in.-lb.

FIG. 12-5.   Variable-speed drive: mechanical. (*Zero-Max Company.*)

Figure 12-6 shows two hydrostatic variable-speed drives which are available in ratings from $\frac{1}{4}$ to 40 horsepower and can be furnished with manual, electrical, hydraulic, or pneumatic controls.

FIG. 12-6.   Variable-speed drives: hydrostatic. (*Renold Crofts, Incorporated.*)

12-4.   Flexible Couplings

The purpose of flexible couplings is to provide a positive drive between two shafts placed end to end while at the same time permitting some angular, radial (parallel), and longitudinal misalignment. In some cases these couplings also provide torsional resilience and end float.

Figure 12-7 shows a *roller-chain-type* flexible coupling consisting essentially of two sprockets connected by a double-roller chain. The split-roller design provides independent roller contact with each sprocket tooth. These couplings will accommodate $\frac{1}{64}$-in. radial misalignment, $1\frac{1}{2}°$ angular misalignment, and a normal amount of end float. Their chief advantages are that they are positive, rugged, simple, compact, and relatively inexpensive. Housings are available to retain the lubricant and keep out dirt.

Figure 12-8 shows a *gear-type* flexible coupling. This type of coupling has a greater torque capacity than most other flexible

FIG. 12-7. Flexible coupling: roller-chain type. (*FMC/Link-Belt.*)

couplings of comparable size, and they provide a somewhat stiffer drive, with good torsional rigidity. The external housing is free-floating and seeks a neutral position in relation to the hubs, accommodating misalignment without imposing added loads on the shafts.

Figure 12-9 shows a *cushion-type* flexible coupling in which the torque is transmitted through a replaceable resilient "starwheel." In addition to being compact and inexpensive, this type of coupling

FIG. 12-8. Flexible coupling: gear type. (*FMC/Link-Belt.*)

Fig. 12-9. Flexible coupling: cushion type. (*Browning Manufacturing Company*.)

offers resilience and vibration dampening and permits an unusual amount of end float.

Figure 12-10 shows a *spring-type* coupling that utilizes a grid-groove design. This type of coupling accommodates shaft misalignment, permits free end float, and provides torsional resilience and

FIG. 12-10. Flexible coupling: spring type. (*The Falk Corporation.*)

shock absorbency. The figure illustrates the action under misalignment and end-float conditions. Notice that under normal load conditions, the grid member bears only at the outer edges of the grooves.
The long span between the points of contact is free to flex under load
variations. The flexibility of this coupling may be altered somewhat
by varying the size of the gap between the hubs.

Figure 12-11a shows a flexible coupling utilizing the classical
Oldham coupling principle. This type of coupling permits a great
amount of parallel misalignment of the shafts, but it lacks the

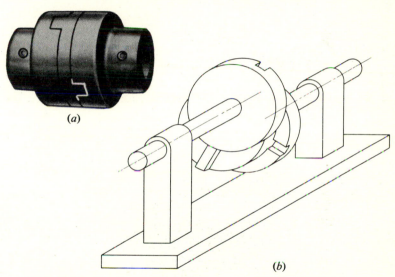

(a)

(b)

FIG. 12-11.   Flexible coupling: Oldham type. (*FMC/Link-Belt.*)

resilience of some of the other types. Figure 12-11b shows how the
Oldham coupling operates. This principle has long been of interest,
because such a coupling is able to transmit constant-velocity motion
regardless of the amount of parallel misalignment.

Figure 12-12 shows a *flexure-plate type* of coupling especially
suited to light-duty applications where zero backlash, torsional
rigidity, quiet operation, and low friction are important. This type

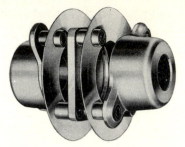

FIG. 12-12.   Flexible coupling:
flexure-disc type. (*Renbrandt, Inc.*)

of coupling accommodates both radial and angular misalignment and
provides a fair amount of resilient end float.

Figure 12-13 shows a somewhat larger, heavier-duty flexure-
plate type of coupling. This coupling is shown partially disassembled
to show one of the two flexible plates and the method of assembly.

FIG. 12-13.   Flexible coupling: flexure-
plate type. (*Formsprag Company.*)

Figure 12-14 shows an unusual *hexical-flexure* coupling which
embodies a curved beam that transmits torque as tension or com-
pression. The flexure is made by precision-machining a helical groove,
or pair of helical grooves, in a solid piece of metal. The remaining
material has a cross section resembling that of a knife blade, and,
just as a knife blade may be flexible, it is extremely difficult to bend

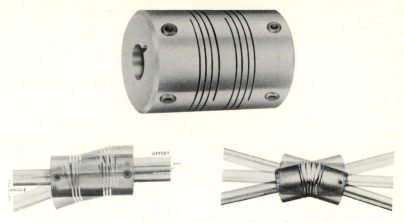

FIG. 12-14. Flexible coupling: helical-flexure type. (*Helical Products Company, Inc.*)

edgewise. This design permits parallel, angular, and skewed misalignments as well as end float. These couplings are available for shafts ranging from $\frac{3}{32}$ in. to $\frac{1}{2}$ in. diameter.

Figure 12-15 shows a *pin-type* flexible coupling utilizing the extremely simple yet rugged system of resilient plugs on one flange mated with clearance holes of the other flange. This design permits a

FIG. 12-15. Flexible coupling: pin type. (*Renold Crofts, Incorporated.*)

fair amount of parallel and angular misalignment and is particularly well suited to applications where a large amount of end float is present.

Figure 12-16a shows how *flexible shafts* can be employed to transmit rotary motion between two shafts that are misaligned or separated by as much as several feet. Such shafts make it possible

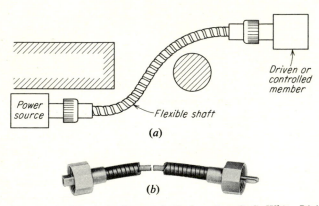

(a)

(b)

FIG. 12-16.   Flexible shaft. (*Pennwalt Corporation, S. S. White Division.*)

to circumvent obstacles or to vary the positions of equipments. As shown in Fig. 12-16b, these shafts typically consist of two end fittings connected by a flexible steel casing that supports and protects a wire-wound drive shaft. Flexible shafts are also available for transmitting *push-pull* motion.

12-5.   Universal Joints

The flexible couplings discussed in the preceding article are designed to accommodate only a small amount of misalignment between two shafts that are *intended* to be coaxial. *Universal joints*, on the other hand, are designed to transmit rotary motion between shafts that are angularly misaligned by as much as 40°.

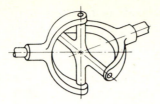

FIG. 12-17. Universal joint:
cardan type (Hooke's joint).

The most common type of universal joint is the *cardan-type* or *Hooke's joint*. As shown in Fig. 12-17, the cardan joint, in its most basic form, consists of three parts—two forks and a center block or spider. Figure 12-18 shows a compact, roller-bearing automotive universal joint of this type. The cardan joint, when used alone, gives nonuniform motion; that is, if the driving fork revolves with constant angular velocity, the angular velocity of the driven fork will vary throughout each revolution. The amount of variation in the output is a function of the angle of misalignment existing between the two forks. If two such joints are used in the same drive system, however, their forks can be oriented so that their velocity variations cancel each other. In the usual automotive application, one joint is used at the transmission and another is used at the rear axle. Since both joints

FIG. 12-18. Automotive universal joint: cardan type. (*Borg-Warner Corporation.*)

operate at about the same angle, uniform velocity is delivered to the wheels.

Figure 12-19 shows a *double-cardan* universal joint, which in reality is two cardan-type joints coupled together with a centering device that divides the angle evenly between the two joints, thus delivering constant velocity.

FIG. 12-19.   Constant-velocity universal joint: double-cardan type. (*Borg-Warner Corporation.*)

Figure 12-20 shows two types of constant-velocity universal joints that utilize steel balls operating in nonconcentric intersecting races in the yokes. The races cause the balls to travel in a plane that always bisects the angle between the driving and the driven shafts. The joint depicted in Fig. 12-20*a* and *b* is a constant-velocity design used primarily in front-drive steering axles. Figure 12-20*c* shows a constant-velocity joint with end-motion capability. This type of joint design is used for sprung-front and rear-drive cars. If the design calls for sliding splines, the drum joint in Fig. 12-20*c* eliminates the need for the sliding member by taking the end motion on the universal joint balls.

The curve in Fig. 12-21*a* shows how the output velocity of a single cardan-type universal joint (bent 30°) fluctuates during one revolution. The curves in Fig. 12-21*b* show how the *magnitude* of the fluctuations increases as the angle between the shafts increases. For angles of less than 10° the velocity variations can usually be neglected, but for angles over 10° the percentage of error becomes important and restricts the use of cardan-type joints to low-speed applications unless they are used in pairs.

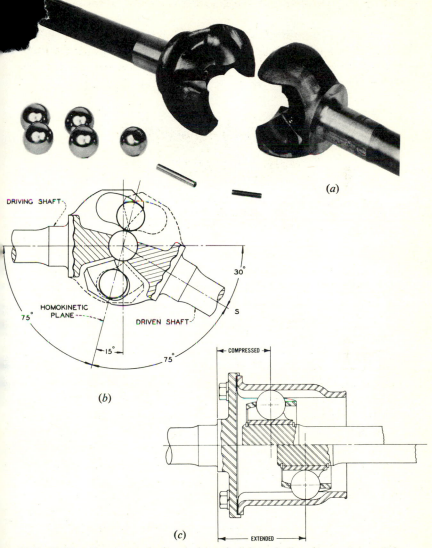

(a)

DRIVING SHAFT

HOMOKINETIC
PLANE

75°

DRIVEN SHAFT

30°

S

15°

75°

(b)

COMPRESSED

EXTENDED

(c)

FIG. 12-20. Constant-velocity universal joint: ball-and-race type. (*The Bendix Corporation, Automotive Control Systems Group*.)

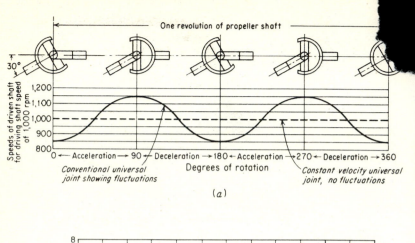

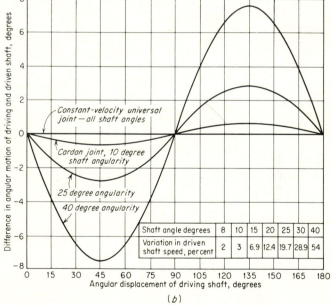

FIG. 12-21.   Velocity curves for cardan-type universal joint. (*The Bendix Corporation, Automotive Control Systems Group.*)

6. Clutches

e word *clutch* is a generic term describing any one of a wide variety
of devices that is capable of causing a machine or mechanism to
become engaged or disengaged. Clutches are of two main types:
positive or jaw clutches, and friction clutches.

Figure 12-22a shows a *positive-jaw* clutch for transmission of
motion in one direction only. The square-jawed clutch in Fig. 12-22b
will transmit motion in either direction, but it is difficult to engage
unless it is made with considerable backlash.

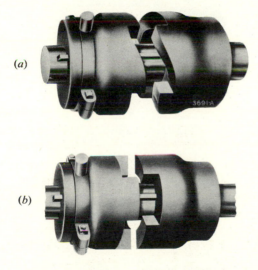

(a)

(b)

FIG. 12-22. Jaw clutches. (*FMC/Link-Belt.*)

Friction clutches are used where engagement must be made
under load or where one member may be moving faster than the other
at the moment of engagement. Typical uses include automobiles and
machine tools. Although usually manually operated by means of a
lever or pedal, friction clutches are often operated electrically,
hydraulically, or pneumatically.

Figure 12-23 shows a *centrifugally operated* friction clutch. As the driving (internal) member revolves slowly, the friction segments slide lightly along the inside surface of the drum-shaped follower. As the angular velocity of the driver is increased, the friction segments are forced harder against the drum and begin driving it.

FIG. 12-23. Centrifugal clutch. (*Formsprag Company.*)

Figure 12-24 shows another form of centrifugal clutch where pivoted, spring-loaded friction shoes are used in place of the loose friction segments. Although more complicated, this type has the advantage that no rubbing occurs until engagement begins.

FIG. 12-24. Centrifugal clutch. (*The Hilliard Corporation.*)

Both of the above centrifugal clutches permit motors or engines to start without load, eliminate the need for flexible couplings, provide overload protection by slipping, and are automatic in their operation.

Figure 12-25 shows a hydraulically operated, multiple-friction-disk clutch that has a high torque capacity and may be operated remotely.

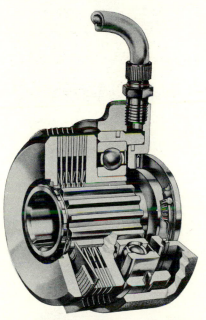

FIG. 12-25. Multiple-friction-disk clutch: hydraulically operated. (*Form-sprag Company.*)

Figure 12-26 shows another type of multiple-friction-disk clutch that is electrically operated.

Figure 12-27 shows a pneumatically operated clutch that consists of multiple-friction shoes actuated by a pneumatic bladder. The simplicity of this design is obvious.

FIG. 12-26. Multiple-friction-disk
clutch: electrically operated. (*Formsprag Company.*)

FIG. 12-27. Multiple-friction-shoe
clutch: pneumatically operated. (*Renold
Crofts, Incorporated.*)

Figure 12-28 shows a *magnetic* clutch based on a relatively new (1948) concept. The clutch consists of two major functional members —stator and rotor. The stator is a stationary coil device. When a direct current is applied to the coil, flux is generated and controlled so that it bridges the powder gaps in the rotor. The powder gaps are partially filled with dry magnetic particles. These particles bind

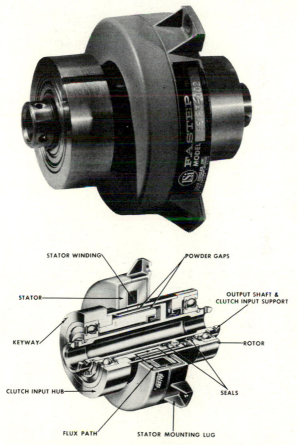

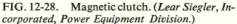

FIG. 12-28.   Magnetic clutch. (*Lear Siegler, Incorporated, Power Equipment Division.*)

together, tying the continuous-rotating input member to the through-shaft output. The amount of torque transmitted is directly proportional to the density of the flux in the powder gap. This allows the clutch to be used as a torque-controlled device in a shearing or slipping mode to a level where the input and output are the same. As in the case of most of the clutches described so far, this clutch can act as a brake as well as a clutch. These clutches have torque capacities ranging from 2 to 70 in.-lb.

OVERRUNNING AND BACKSTOPPING CLUTCHES

Figure 12-29*a* shows a *sprag-type*[1] clutch that can be used (*a*) as an overrunning clutch whereby the driver automatically disengages when the driven member exceeds the driver speed, (*b*) as a backstopping clutch, to prevent the backward rotation of a shaft, or (*c*) as an indexing device providing silent, stepless ratchet action. The wedging action of the individual sprags is illustrated in Fig. 12-29*b*. If the outer ring represents the driven member and the inner ring the driver, it is evident that, if the driver is turned to the right, the sprags will wedge tightly between the two rings and drive the outer ring. If the driver is turned to the left, the sprags will slip.

Figure 12-30 shows a *roller-type* clutch whose operating characteristics are the same as the sprag-type clutch, except that rollers and cam surfaces produce the wedging action rather than sprags. The principle involved is the same as that of the *silent* ratchet shown schematically in Fig. 12-38.

Figure 12-31 shows a *friction-type backstop* which permits rotation in one direction but not the other. A backstop of this type is often used in elevators or inclined conveyors where a safety device is required to prevent the mechanism from reversing in the event of power failure. This particular type of backstop is somewhat less versatile than the sprag-type or roller-type clutches, but it has the advantage of being simple, inexpensive, and smoother-acting.

[1] For a good discussion of sprag-type clutches see W. T. Cherry, Application of Sprag-type Overrunning Clutches, *ASME Paper No. 57-A-165.* Available from The American Society of Mechanical Engineers, 29 West 39th St., New York, N.Y.

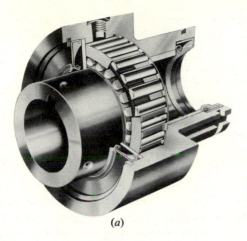

(a)

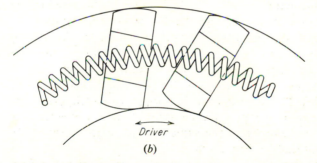

Follower

Driver

(b)

FIG. 12-29.   Sprag-type clutch. (*Formsprag Company.*)

SINGLE-REVOLUTION CLUTCHES

Many machines are driven by continuously running motors and yet perform cyclical or intermittent operations on command. Machines of this type include shears, punch presses, riveters, staplers, and intermittent conveyors. This is often accomplished with a single-revolution clutch, that is, a clutch which when activated makes a complete revolution and then disengages itself and allows the motor

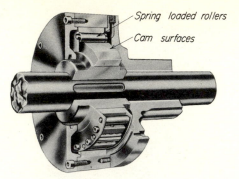

FIG. 12-30. Roller-type clutch. (*The Hilliard Corporation.*)

FIG. 12-31. Friction-type backstop. (*FMC/Link-Belt.*)

to run free until the next cycle is called for. Figure 12-32 shows a commercially available single-revolution clutch whose driving action is accomplished through a set of rollers wedged between an outer race and an inner cam race. The trip lever causes the cam to wedge the rollers in the same manner as that of the roller-type overrunning

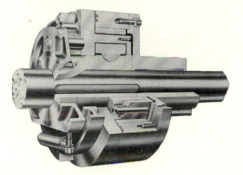

FIG. 12-32. Single-revolution clutch. (*The Hilliard Corporation.*)

clutch. When one revolution is completed, the cam permits the rollers to release. If extremely accurate operation is desired, a positive stop mechanism is available to eliminate any possible overrun error. These clutches can also be furnished with multiple stopping points.

HYDRAULIC COUPLINGS

Figure 12-33 shows a typical fluid coupling utilizing two turbine-like paddle wheels facing each other in an enclosure filled with hydraulic fluid. These couplings are becoming popular as replacements for friction clutches. They provide practically no-load starting and idling for motors and engines, smooth load pickup, overload protection, and a soft, cushioned drive usually called *fluid drive*. They are virtually wearproof and maintenance-free. Their disadvantages include high initial cost, bulk, and reduced efficiency. Whereas friction

FIG. 12-33.   Hydraulic coupling. (*Renold Crofts, Incorporated.*)

clutches stop slipping once the load is picked up, hydraulic couplings always slip a little, so a certain amount of power is consumed in heating up the fluid.

12-7.   Intermittent-motion Drives

One of the oldest devices for producing intermittent motion is the *ratchet*, consisting of a *ratchet wheel* and *pawl*, as shown in Fig. 12-34. As the lever is raised, the pawl drives the wheel; as the lever is lowered, the pawl slips over the ratchet teeth. Thus, a reciprocating motion of the lever produces an intermittent rotary motion of the ratchet wheel. A second pawl, shown in dashed lines, is often included to prevent the ratchet wheel from being dragged backward.

The *double-pawl ratchet* shown in Fig. 12-35 has the same action as a single-pawl ratchet having twice the number of teeth, yet it has the advantage of the larger, stronger teeth.

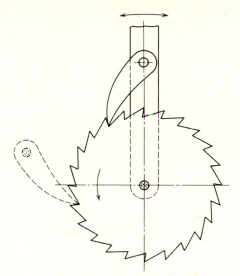

FIG. 12-34. Single-pawl ratchet.

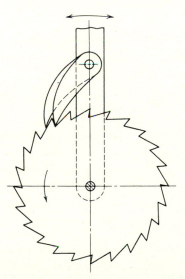

FIG. 12-35. Double-pawl ratchet.

Figure 12-36 shows a *double-acting ratchet*, whereby the ratchet wheel is driven on both forward and backward motions of the lever.

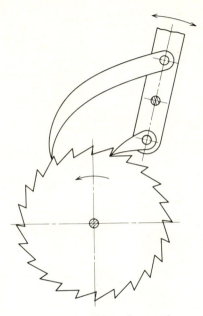

FIG. 12-36.   Double-acting ratchet.

Figure 12-37 shows two forms of *reversing* ratches that enable the ratchet wheel to be driven in either direction. This is frequently necessary in the case of table feeds on machine tools.

Silent ratchets are ratchets that have no teeth but depend instead on a wedging together of smooth surfaces. Kinematically, they perform like a ratchet having an infinite number of teeth on the ratchet wheel, and there is none of the familiar clicking of the pawl. The sprag-type and roller-type clutches discussed in the preceding article (Figs. 12-29 and 12-30) are in effect silent ratchets. Figure 12-38 shows two common forms of the silent ratchet. The ball-type silent ratchet often has springs behind the balls to ensure a positive wedging action.

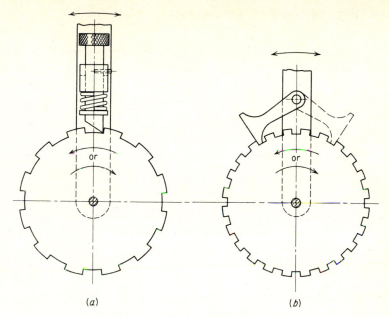

(a)

(b)

FIG. 12-37.   Reversing-pawl ratchets.

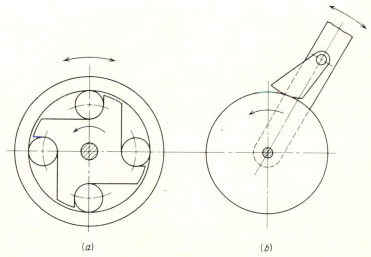

(a)

(b)

FIG. 12-38.   Silent ratchets.

Figure 12-39 shows a typical ratchet-drive arrangement used for stock feed rolls on punch presses, where the *throw* of the driving crank is adjustable for varying the amount of ratchet-wheel rotation per cycle. Ratchets may also be driven by cams or eccentrics.

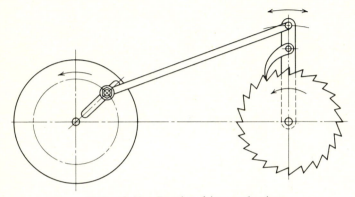

FIG. 12-39.   Ratchet-drive mechanism.

The *Geneva mechanism*, shown in Fig. 12-40, is a popular device for obtaining intermittent motion from a constant-velocity driver. It is also popular as an indexing device for machine tools such as turret lathes and screw machines. In the particular mechanism shown, the driven member rotates one-fifth revolution for every revolution of the driver. In Fig. 12-40*a*, the circular portion of the driver prevents the follower from turning. In Fig. 12-40*b*, the drive pin is just about to begin driving the follower, and, simultaneously, the circular portion on the crank is just releasing the follower. In Fig. 12-40*c*, the pin is driving the follower, and the circular portion of the crank is entirely free of the follower.

Since Geneva mechanisms are widely used in special machinery and in machine tools and since these machines usually involve rather heavy members, it is often necessary to analyze the velocities and accelerations imposed on the driven member. The velocity analysis of a Geneva mechanism involves the method explained in Art. 6-10; the acceleration analysis involves the method explained in Example 7-8.

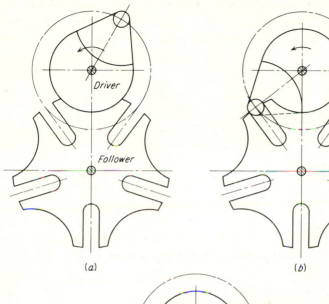

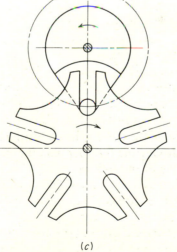

FIG. 12-40.   Geneva mechanism.

bibliography

Beggs, J. S.: *Advanced Mechanism*. New York: The Macmillan Company, 1966.

Cowie, A.: *Kinematics and Design of Mechanisms*. Scranton, Pa.: International Textbook Company, 1961.

Dudley, D. W. (ed.): *Gear Handbook*. New York: McGraw-Hill Book Company, 1962.

Faires, V. M.: *Kinematics*. New York: McGraw-Hill Book Company, 1959.

Faires, V. M., and Keown, R. M.: *Mechanism*, 5th ed. New York: McGraw-Hill Book Company, 1960.

Hain, K.: *Applied Kinematics*, 2d ed. New York: McGraw-Hill Book Company, 1967.

Hall, A. S., Jr.: *Kinematics and Linkage Design*. Englewood Cliffs, N.J.: Prentice-Hall, Inc., 1961.

Ham, C. W., Crane, E. J., and Rogers, W. L.: *Mechanics of Machinery*, 4th ed. New York: McGraw-Hill Book Company, 1958.

Harrisberger, L.: *Mechanization of Motion*. New York: John Wiley & Sons, Inc., 1961.

Hartenberg, R. S., and Denavit, J.: *Kinematic Synthesis of Linkages*. New York: McGraw-Hill Book Company, 1964.

Hinkle, R. T.: *Kinematics of Machines*, 2d ed. Englewood Cliffs, N.J.: Prentice-Hall, Inc., 1960.

Hirschhorn, J.: *Kinematics and Dynamics of Plane Mechanisms*. New York: McGraw-Hill Book Company, 1962.

Huckert, J.: *Analytical Kinematics of Plane Motion Mechanisms*. New York: The Macmillan Company, 1958.

Lent, D.: *Analysis and Design of Mechanisms*. Englewood Cliffs, N.J.: Prentice-Hall, Inc., 1961.

Martin, G. H.: *Kinematics and Dynamics of Machines*. New York: McGraw-Hill Book Company, 1969.

Maxwell, R. L.: *Kinematics and Dynamics of Machinery*. Englewood Cliffs, N.J.: Prentice-Hall, Inc., 1960.

Phelan, R. M.: *Fundamentals of Mechanical Design*, 3d ed. New York: McGraw-Hill Book Company, 1970.

Rothbart, H. A.: *Cams*. New York: John Wiley & Sons, Inc., 1956.

Sahag, L. M.: *Kinematics of Machines*, revised. New York: The Ronald Press Company, 1952.

Shigley, J. E.: *Kinematic Analysis of Mechanisms*. New York: McGraw-Hill Book Company, 1969.

Spotts, M. F.: *Mechanical Design Analysis*. Englewood Cliffs, N.J.: Prentice-Hall, Inc., 1964.

Tao, D.C.: *Fundamentals of Applied Kinematics*. Reading, Mass.: Addison-Wesley Publishing Company, 1967.

Tuttle, S. B.: *Mechanisms for Engineering Design*. New York: John Wiley & Sons, Inc., 1967.

Tyson, H. N.: *Kinematics*. New York: John Wiley & Sons, Inc., 1966.

Zimmerman, J. R.: *Elementary Kinematics of Mechanisms*. New York: John Wiley & Sons, Inc., 1962.

index